Insect Magic

Insect Magic

Photographs by Kjell B. Sandved
Text by Michael G. Emsley

A Studio Book · The Viking Press · New York

Text Copyright © Viking Penguin Inc., 1978
Color illustrations Copyright in all countries of the
International Copyright Union by Viking Penguin Inc., 1978

First published in 1978 by The Viking Press
625 Madison Avenue, New York, N.Y. 10022

Published simultaneously in Canada by
Penguin Books Canada Limited

Library of Congress Cataloging in Publication Data
Sandved, Kjell Bloch, 1922–
    Insect magic.
    (A Studio book)
    Bibliography: p.
    Includes index.
    1. Insects.   I. Emsley, Michael G.
II. Title.
QL467.S33   595.7   78-2216
ISBN 0-670-39908-6

Text printed in the United States of America

Color illustrations printed in Japan

Contents

Acknowledgments

First of all I wish to thank my good friend and helper Barbara Bedette for her constant assistance and encouragement in the making of this book. A special debt is owed to my many colleagues and friends in the National Museum of Natural History, particularly to Dr. Edward S. Ayensu for his unfailing support and good advice through the years.

I extend my gratitude to the publishers, Viking Penguin, for this, my third book with them, and particularly to Olga Zaferatos for her patience and suggestions.

A special thanks to my many helpers in the field, who enabled me to photograph many rare specimens in faraway places. Some of the species have never been photographed before, and on occasion my photographs were of species new to science. I owe an immeasurable debt of gratitude to the following persons: Dr. Donald R. Davis, Dr. W. Donald Duckworth, William D. Field, Dr. Richard C. Froeschner, Francis M. Greenwell, Thomas R. Harney, Dr. Porter M. Kier, Jack F. Marquardt, Dr. Robert W. Read, and Dr. Paul J. Spangler, *Smithsonian Institution*; Dr. Amada A. Reimer and Dr. Nicolas Smythe, *Smithsonian Tropical Research Institute, Canal Zone*; Dr. Donald M. Anderson, Dr. Robert D. Gordon, Dr. Ronald W. Hodges, Dr. Lloyd V. Knutson, Dr. Curtis W. Sabrosky, Rose Ella Warner Spilman, T. J. Spilman, George C. Steyskal, and Dr. Edward L. Todd, *USDA, Washington, D.C.*; Dr. Soernatono Adisoemarto, *Lembaga Biologi Nasional, Indonesia*; Dr. Graziela M. Barron, *Jardim Botanico do Rio de Janeiro*; Drs. Roger and Ookeow Beaver, *Chian Mai University, Thailand*; Professor Clifford O. Berg, *Cornell University*, Dr. Roberto Burle-Marx, *Rio de Janeiro*; Professor Charles E. Cutress, *University of Puerto Rico*; Professor Howard E. Evans, *Colorado State University*; Dr. Graham Bell Fairchild, *Panama City*; Dr. J. Linsley Gressitt, *Wau Ecology Institute, New Guinea*; Dr. M. P. Harris, *Galapagos Islands*; Professor Kenneth P. Lamb, *University of Papua New Guinea*; Professor Martin Naumann and Dr. Dennis Leston, *University of Connecticut*; Dr. João Murca Pires, *IPEAN, Belém, Brazil*; Dr. Ivan Polunin, *University of Singapore*; Professor Diomedes Quintero, *Universidad de Panama*; Judy Rodden Schnedl, *Washington, D. C.*; Professor Margot Schumm, *Montgomery Community College, Maryland*; Professor Laura Schuster, *Universidad A. de la Selva, Peru*; Dr. Alcides R. Teixeira, *Instituto de Botanico, São Paulo*; Richard Thacker, *University of Maryland*; Professor Roman Vishniac, *Yeshiva University*; Professor Thomas J. Walker, *University of Florida*; David Wapinski, *University of Virginia*; Kirsten Wegener-Kofoed, *Copenhagen, Denmark*; J. S. Womersley, *Lae Botanical Gardens, New Guinea*; Choo See Yan Brothers, *Cameron Highlands, Malaysia*; Professor Fernandez Yepez, *University of Maracay, Venezuela*.

K.B.S.

Foreword

I count myself very fortunate to have joined the Imperial College of Tropical Agriculture in Trinidad before the late Professor Kirkpatrick retired in 1960, for KP's enthusiasm and prodigious knowledge provided me with the motivation I sorely needed as a young man.

KP came from a scholarly family and one of his charming skills was the impromptu creation of "clerihews" about his colleagues. A clerihew, named after Edmund Clerihew Bentley, is a pair of couplets of unequal length which rhyme aa/bb, like this:

> KP's life was in the tropics
> Digging into diverse topics.
> On insects bent,
> A real gent.

Thank you, KP, for pointing me in the right direction.

Michael G. Emsley
George Mason University
Fairfax, Virginia

Introduction

Dree, dree . . . grrr, grrr . . . tch, tch, tch. . . . The chirping of crickets, the singing of grasshoppers, the crescendo and diminuendo of cicadas. A scene from a meadow along the Potomac River? No! It's all part of the new Insect Zoo at the Smithsonian Institution's National Museum of Natural History. The Smithsonian has always explored new ways to promote "the increase and diffusion of knowledge among men," and Dr. Porter M. Kier, the director of the National Museum of Natural History, took an innovative step in 1975 when he created an exhibit hall devoted solely to live insects and their relatives. Local insects and also some exotic ones are exhibited here in enclosures that resemble the insects' natural habitats.

The Smithsonian Insect Zoo offers thought-provoking insight into the lives of insects and gives the visitor a sense of participation and personal discovery. It helps the individual to gain an understanding of biological principles in action, and, it is hoped, improves the urban visitor's sometimes stereotyped image of insects as ugly bugs.

Through their behavior the insects in the Zoo relate part of their own story, which is infinitely more dramatic and informative than what is revealed by labels on cabinets containing insects mounted on pins. Visitors are surprised by the unusual appearance and unpredictable actions of many of the insects, and are anxious to learn something about their lives. There is so much going on in the various enclosures that some visitors just pull up a chair and sit for an hour or more in front of one of the exhibits, watching the actors in nature's own miniature theater. Feeding demonstrations usually stir great delight among the visitors, who may volunteer to feed the insects.

Among the most interesting social insects in the Zoo are the leaf cutters and fungus ants. Like bees and termites, these ants have a strict caste system, with a queen, reproductives, and workers and soldiers which act as scouts, nursery attendants, guards, and garbage collectors.

Each day fresh flowers and leaves are placed in a large enclosure connected to the ant nest by a transparent tube. Scout ants discover the food source and make scented trails to the nest, where they recruit workers. At the food source the workers cut neat pieces of petals or leaves and carry them back along the scented trail. The cut pieces are often larger than the ants and would correspond to our carrying a Masonite board, measuring eight by ten feet, between our teeth! To this feat may be added the extra burden of carrying a free-riding police ant whose function supposedly is to attack enemies. The cut petals and leaves are carried into the growth chamber, where the temperature and humidity are controlled by the ants themselves. Nursery-attendant ants chew each piece to protect it from growth of unwanted fungi. Only the ants' own fungus is allowed to grow on the small cut pieces. This nest-grown fungus is rich in protein, and forms the basic food of these ants.

As in any society, including our own, there are always inedible left-over scraps, so some of the workers act as trash collectors. They pick up the refuse with their mandibles and carry it out of the nest. In some species a remarkable behavior pattern has evolved through the ages: the trash carriers do not throw away the garbage indiscriminately as human litter-bugs sometimes do, but march a considerable distance away from the nest, climb up a tree and out on a branch, where at one point, and only one point, they let the little shriveled ball of trash fall. It literally rains to the ground and eventually forms a huge mound. Once when filming leafcutting ants in Panama, I could hardly believe my eyes when I came upon trash carriers that went way out on a limb overhanging a little stream to plop their load into the running water.

No insect zoo would be complete without a display of bees. In 1926 the Smithsonian began a display of honeybees, which was frequently visited by the First Lady, Mrs. Calvin Coolidge, who often brought her friends to see it. One day the bees swarmed and settled—where else?—in the back yard of the White House. The bees were soon evicted, and Mrs. Coolidge's interest in them cooled.

For years after that incident bees were not exhibited at the Smithsonian. When, later, another colony was installed, the bees swarmed to a nearby air conditioner; later still, in 1963, the resident colony fell victim to a "silent spring" because of heavy doses of insecticide sprayed on the Mall.

10

The present colony in the Insect Zoo is seemingly thriving, and is busy gathering soda pop spilled by the Mall's tourists. The bees are exhibited behind glass so that all activities can readily be seen. A plastic tube connects the hive to the outside through a hole in an adjacent window.

The workers are able to sense the stages of hive construction and relieve each other at intervals of about half a minute, each building part of the cell. Each new cell wall is built at exactly 120 degrees to the next cell wall, utilizing minimum materials with maximum strength. Where numerous cells touch each other the resultant structures are hexagons, as in corn on the cob. A four-year-old, watching the hive being built, turned to his mother and exclaimed, "They're making corn!" A good observer.

It has been a particular pleasure for me to work on this book, since it incorporates many ideas of my first project in the Smithsonian Institution: writing a comprehensive proposal for a new Hall of Insects for the National Museum of Natural History. This project later evolved into the Insect Zoo.

Nowhere in the animal kingdom do we find such diversity of behavior and such pleasing color play as in the world of insects. Among the most admired insects are the butterflies, one of nature's most beautiful gifts to mankind. Every art form has used their colors and designs as embellishment. Looking long enough, one can even find the entire alphabet, as well as numerals, in the form of scale designs in their wings. The beauty in a butterfly's wing speaks a universal language. The same can be said for countless beetles and other insects. "Natural beauty in its most profound sense," said Dante, "is the highest form of beauty."

Many ask, "Where can we find interesting insects to observe and photograph in nature?" The answer is "Almost everywhere." From the tropical rain forests to the deserts, in oceans and in ponds, and in all countries of the world, as the examples in this book illustrate. But we do not need to travel to exotic places to experience the marvelous kingdom of animals and plants. How many of us on a summer's day in a field, or at leisure in our own back yard, pause and observe the beauty and variety of the many plants and animals there? Many of them require a magnifying lens to be perceived. The scientist Louis Agassiz, who found great drama in the everyday life around him, once said, "I spent the summer traveling. I got halfway across my back yard."

It is true that in spite of the fact that insects are the most numerous animals on earth, few people notice more than a very small percentage of the insect species that surround them. Why are insects so numerous? Or should we rather ask: why are they so successful?

As life evolved on earth the insects excelled in their ability to invade new niches and to exploit all possible, and some seemingly impossible, ecological microhabitats. Every habitat, every niche is a complex of insect competition, where each group, each species, evolved many distinctive characteristics that have enabled them to adapt to their particular environment. The diversity of their habitats, feeding techniques, life histories, and shapes and colors is the reason for their success. The successful were winners in the evolutionary game. The unsuccessful species are now extinct.

In the past few decades much has been learned about predator-prey relationships. Just think of the visual acuity and learning ability of birds, or the protective and warning coloration of insects. Think of the evolution of sonar, used by bats for navigation and location of prey at night, or the warning signals emitted by some arctiid moths to avoid being caught. Yet relatively little is known of the actual life cycle and behavior patterns of tropical insects even though they are well represented in our museum collections. So many interesting facts and so much important data remain to be learned. This challenge awaits the future student.

The marvel about insects' behavior is that it is genetically coded in a brain the size of a pinhead. There lie the biological clocks, the stimulators, the inhibitors, and the promoters for dictating strategies in the evolutionary game. DNA, the dictating molecules in the brain of an insect and in man, has little regard for whether those molecules have to produce a tiny beetle or a curator of beetles in a museum. The exploration of this multicolored biological jungle has scarcely begun.

Why is it so important that we should continue to learn more about the behavior of insects? "Man has only begun to document the intricate web of relationships among species on this planet," says Smithsonian Secretary S. Dillon Ripley, "and has yet to comprehend or appreciate his own place in the natural world."

As man searches for an understanding of his own kind, he finds many aspects of his behavior biologically rooted in his dim past, from primordial protozoans to primates. Animals in nature are confronted with the same basic problems which confront man, but animals manage to solve them in a number of wonderfully different ways. Awareness of and familiarity with these strategies induce a young naturalist to capture a butterfly with the camera lens, rather than with a net. It might even bend *Homo sapiens'* inexorable lust for game hunting and shooting, so that one day he might—I say *might*—shoot with a camera rather than a gun.

If not, the tenacity and durability of insects will eventually be more than a few steps ahead of man. This brings to mind entomologist W. J. Holland's poetic vision of who would inherit the earth:

> When the moon shall have faded out from the sky, and the sun shall shine at noonday a dull cherry-red and the seas shall be frozen over, and the icecap shall have crept downward to the equator from either pole, and no keels shall cut the waters, nor wheels turn in mills, when all cities shall have long been dead and crumbled into dust, and all life shall be on the very last verge of extinction on this globe; then, on a bit of lichen, growing on the bald rocks beside the eternal snows of Panama, shall be seated a tiny insect, preening its antennae in the glow of the worn-out sun, representing the sole survival of animal life on this our earth—a melancholy bug.

—Kjell B. Sandved

National Museum of Natural History
Smithsonian Institution
Washington, D.C.

I. The Beginnings

In the fall of the year in which I turned thirteen, my family moved from Bedford to a small English country town about twelve miles away, and for the next five years I was to travel to school by bus. There were rather few children of my own age engaged in this daily pilgrimage, but there was one boy, two years my senior, who was destined to affect my life profoundly.

Simon Barnes was old enough to command my respect, yet sufficiently near my own age for him to find me tolerable. I was enormously impressed by his skill at the piano, his racing bicycle, and his collection of butterflies. That winter I persuaded my decidedly unwealthy father to buy a piano, which I proceeded to pound incessantly. During moments of digital relaxation I drilled holes in the frame of my "sit-up-and-beg" touring bicycle to lighten it and make it look more racy. Then came the spring and the butterflies. . . . Simon and I were inseparable, cycling dozens of miles each day in search of good localities where we could find new specimens to add to our collections. With only sixty-eight resident species of butterfly known in England, we soon had the majority of them checked off. Within a few months Simon had remolded me in his image, a goal which my father, who was an amateur microscopist, had failed to achieve in more than a decade.

After school Simon became a surveyor and I a professional entomologist. Would I have chosen a career with insects if I had not met Simon Barnes? Almost certainly not. It is quite extraordinary, and somewhat disturbing, how chance relationships so easily change the course of one's life. I am convinced that the friends you make when you are young are the most influential people you ever meet. Parents with young children, take heed!

It was during this early period of butterfly mania that I was able to cajole my father into taking me to Wicken Fen, the last English stronghold of the swallowtail butterfly *(Papilio machaon)*. The butterfly was even then

15

protected by law, and the wild population was periodically enhanced by the release of laboratory-reared insects. Collecting the swallowtail was prohibited, but I secretly intended to make a capture if no one was looking. Shortly after we arrived, I was walking along a narrow path through the rank grass when I spotted a kind of butterfly that I had never seen before. As it flew erratically toward me, I swept it into my net and immediately recognized it as a rare variety of the comma butterfly *(Polygonia c-album)*. I had just ended its life with a quick pinch of the thorax when I felt the closing presence of two full-grown men, who I rapidly gathered had been pursuing the butterfly that I had caught. I ran off as fast as I could and spent the remainder of the afternoon studying pond life with my father. He was highly flattered.

The incident went forgotten for nearly ten years until, as a young and insecure undergraduate, I was attending my first Royal Entomological Society of London tea. For about an hour prior to each of the monthly lectures the society members meet in the library and consume inordinate quantities of tea and glutinous pastries while discussing their latest research. It was there, while consuming an oozy bun and trying hard to look as if I belonged, that I overheard a conversation between two distinctly mature gentlemen that stopped my jaws in mid-bite. One man, who I now believe was a most eminent geneticist, was recounting how, many years previously, a resident naturalist at Wicken Fen had reported sighting an aberrant variety of the comma butterfly for the first time in many, many years. The geneticist and a colleague had gone to the Fen for the express purpose of collecting live specimens, in order to determine whether the form was a genetic mutation or had been environmentally induced. "Damn it," the man said, "you won't believe it. We had just found one when this wretched little whippersnapper rushed out of the bushes and killed it! If I had caught him, I would have wrung his ruddy neck! We never saw another one that day." I eased myself quietly away lest I should be recognized, desperately hoping I would never meet this man face to face at a job interview.

II. Joy and Satisfaction

The relationship most citizens of the Western world have with nature is characterized by their attitude toward insects. "Ugh! Yuk! There's a bug. Squash it!" Is this, one wonders, the reason the English call thick rubber-soled shoes "beetle crushers"? It seems that ladybugs and butterflies are the only insects that escape the blanket assertion that all bugs are beastly.

Universally hostility and fear are the products of ignorance, and our antagonism toward insects is no exception. One cannot deny that insects are a nuisance when their bites become sore, and a threat when they transmit disease, but, viewed dispassionately, even these noxious insects are beautiful.

It saddens me that so many people deny themselves the simple pleasure of appreciating the exquisite elegance of even the more common insects. It is not only the superb sculpturing of their cuticles but the extraordinary complexity of their lives that make insects such a fascinating source of enjoyment. It takes only a moment with a hand lens to discover a whole new world of design and beauty. The pictures in this book just begin to manifest the excitement of discovering the patterns that are the daily fare of the entomologist.

To the question, "Why are insects here at all?" different people have different answers. As an evolutionist, I would answer that they are the surviving products of a natural system which rigorously selects animals and plants for their "goodness-of-fit" in the environment. From having studied their fossilized remains, we know that insects have inhabited the earth for nearly four hundred million years, though they do not seem to have become prominent until about two hundred and fifty million years ago, from which time we have remnants of insects that are similar to modern mayflies, dragonflies, beetles, and cockroaches. However, the massive proliferation of insects toward their present abundance and diversity had to await the

17

evolution of the flowering plants, with which they now have an intimate relationship as pollinators and parasites.

Pests are pests only because they compete directly with man. In a man-free environment the diversity of plant and animal life is so great that no single species can achieve pestilential proportions, for as one organism becomes more numerous, so it becomes food for another and its population declines. Animals disperse and occupy any niche in which they can make a living; thus, natural communities are a dense web of closely interlocking niches, each tenuously occupied by a particular species. Change in the weather, or the introduction of a new animal, alters the precarious balance of the organisms in the community and the species' composition shifts as a new balance is struck.

Today, as we cut forests, burn brush, change water courses, and extend our urban sprawl, man's intervention is the most serious disturbing force. But perhaps more than in any other single way, man has disturbed Nature by planting vast acreages with single species of plants such as corn, wheat, rice, peanuts, and other commercial crops. We should hardly be surprised to find that in these unique environments there have arisen uniquely large populations of insects, for wherever food is abundant some animal will exploit it. The insects, because of their mobility and powerful reproductive rates, are the first organisms to become numerous enough to be noticed. To an entomologist, these facets of modern life are just a part of the fascination of understanding insects.

We find insects everywhere we look, sometimes even in cigarettes and breakfast cereal, yet the utilization of insects as human food has been largely neglected, although in Africa locusts are not regarded as without value and are regularly eaten. Some years ago, in a moment of curiosity, I sampled a fresh grasshopper and was relieved to find that it tasted only of cooking-fat and brown paper. I first ate a female, chauvinistically perhaps, and my waning enthusiasm did not permit a second trial. Maybe males are a gourmet's delight!

In this age of chronic protein deficiency among the poorer peoples, it is encouraging to learn that the resources provided by insects are not being totally overlooked. In a recent scientific article, Kenneth Ruddle records the insect-eating habits of the Yukpa Indians on the Venezuela-Colombia border. The Yukpa claim that the rhinoceros beetle is particularly delicious, not because of its flavor but because "it contains a lot of meat." Ruddle explains, "In the culinary preparation of the adult form, most people remove the legs, wings, thorax, and head. The abdomen, skewered on a small stick, is then thrust briefly into the fire to toast lightly. Some, mostly the children it seems, prefer to eat these beetles in a raw state. . . . The elytra [wing cases] and hind wings . . . are sometimes used as necklace ornaments, and the femur . . . is used as a bead separator."

Though pickled beetles, which allegedly taste of almonds, were sold by street venders in China some fifty years ago, in the United States we are conditioned to regard insects as edible only when coated in chocolate and served as cocktail delicacies. I wonder if sales would decline if more people knew that honey had already been swallowed and regurgitated by a bug!

My appreciation of insects as a source of pleasure does nothing to prevent me from killing specimens for study, yet I would shrink from being commissioned to kill a chimpanzee, elephant, or dog. Wherein lies the difference? I confess I do not know. Peter Singer, writing in *Animal Liberation: A New Ethic for Our Treatment of Animals*, extrapolates from the familiar but uncomfortable prejudices of racism and sexism to a new and poorly recognized prejudice, that of "species-ism." As *Homo sapiens*, have we elevated ourselves to a level of self-importance that fails to recognize that other organisms might also have rights? However, while it is sanctimoniously easy to support the "right to life" of the larger animals, my devotion to insects' rights becomes instantly attenuated when a mosquito probes my arm.

The real thrill of entomology, as a hobby or profession, comes from the enrichment of commonplace events. An automobile radiator becomes a minimuseum, household insects take on a measure of companionship, and under every stone and rotting log are potential joy and satisfaction.

III. Insect Intimacies

I had absentmindedly picked at the pimple on my wrist for several days before examining it closely. Eventually I looked carefully and was horrified at what I saw. I had seen round pinhole-sized craters previously but they had always been on dogs and served as breathing holes for maggots feeding just below the skin. Having just returned from a collecting trip on the slopes of the eastern Andes, I immediately realized that I had become the victim of the human bot fly *(Dermatobia hominis)*. This fly is amazing, for it finds and captures a blood-sucking insect, such as a female mosquito, and cements one or more eggs to her body. During one of the mosquito's later blood meals, triggered either by the warmth or odor of the host, the eggs hatch and the tiny maggots wriggle out of their eggshells. Most often, the maggots drop harmlessly to the ground and perish, but sometimes, as on my wrist, at least one maggot alights on the skin and burrows in to feed on the underlying tissues.

After I had overcome my initial surprise, I remembered a microscope slide I had seen at college. Under the glass cover slip was the evacuated skin of a large *Dermatobia* maggot with a label reading "removed from the skin of OWR" (O. W. Richards was my senior professor). I was impressed. So, egotistically, I determined that the University of the West Indies should have a souvenir from me.

To cut off the air supply, I piled transparent grease over the cavity, and, as expected, a maggot reached out to breathe. It was disappointingly small, so, with trepidation, I decided to keep it until it would make a more impressive specimen. However, after four or five days I became progressively concerned over what the grub might be eating, and I began to imagine it consuming blood vessels, muscles, nerves, and other important parts. My determination failed, and I set about extracting it. By piling the grease very high, I coaxed the maggot out until only its terminal hooks were gripping the rim of its chamber. In this extended position I grasped it gently with a pair of forceps and carefully pulled it out. It was with some relief that I found only a single specimen no more than half an inch long, but fully worthy of being prepared and appropriately labeled "Removed from the skin of MGE."

In West Africa similar flies used to lay their eggs on laundry drying in the sun, the eggs hatching when the clothing was worn. These tumbu flies (Cordylobia anthropophaga) were greatly feared by colonial white mothers because their babies' diapers were the flies' favorite targets.

These unpleasant experiences are quickly overwhelmed by the consummate pleasure of discovering the intimacies of tropical insect life. For example, many tropical plants grow on the branches of trees, where, without contact with the soil, their principal difficulty is gathering nutrients. Some species of these plants have solved their nutritional problem by developing a specially expanded region of their stem into a labyrinth of tunnels, the ends of which are absorptive, whereas the principal walls are not. Ants eagerly occupy the tunnels as a nesting site and, in exchange for the accommodation provided, they dump their alimentary and kitchen waste in the absorptive tunnel endings where, it seems, the nutrients are taken in by the plant. In some cases, the ants even protect the plant from attack by other insects and collect and sow the plant's seeds in sites suitable for germination.

It is not only in the tropics that one finds such exotic relationships, for one of the most amazing associations is between an ant (Lasius) and a root-sucking aphid (Aphis maidiradicis) in the southern United States. During the winter the ants nurture the eggs of the aphids in their underground nests, and then, with the arrival of spring, the newly hatched aphids are carried in the jaws of the ants onto the roots of nearby weeds. Later, with the vigorous growth of local farmer's corn, the ants round up the aphids and transfer them to the roots of the corn, where, during the summer, their numbers can reach pestilential proportions. In the fall the ants collect the eggs laid by the aphids and carry them back to their nests to survive the winter. Do the ants profit from this relationship? As one might expect, altruism seems as rare in nature as it does in man, and there is indeed a reward. When an ant strokes an aphid with its antennae, the aphid responds with the production of a sweet syrup called "honey-dew," a secretion which

22

is much sought after by many species of ants and seems to form the foundation of their diverse relationships. It is not surprising, then, that aphids are sometimes known as "ant-cows." There are even ants which bite off their aphids' wings to prevent them from flying away.

Insects occur abundantly everywhere except in the sea, and they are absent from the oceans only because, evolutionarily, the ancestors of the crabs and shrimps arrived there first. However, a few insects have successfully invaded tidal pools and sandy shores, and water striders may be found walking on the surface of the sea thousands of miles from land. Insects have been found on the snow of the polar caps and the peaks of high mountains, and have even been collected in aerial nets towed behind aircraft flying at altitudes above ten thousand feet.

In terms of food, there is hardly any material which some insect does not eat. Some beetles even eat through lead cables, though why they do it is far from clear. It has been suggested that they are attracted by the electric field created by the current passing through the wires, but the experimental evidence to support that idea is doubtful at best. The blood of animals is drawn by flies, fleas, lice, and bedbugs, while other lice nibble away at hair and feathers. There is hardly a part of any plant that is not eaten by insects, although many plants have evolved complex chemicals to deter insects and other herbivores from considering them as a food source. Some of these plant protectors are poisons, while others inhibit the digestive processes of the insects or render them sterile; either way the insect numbers are eventually reduced. But for every protective device there is usually one or more insect species that has evolved a counter strategy; for instance, while nicotine is a very effective insecticide, the cigarette beetle (*Lasioderma serricorne*) is a worldwide pest of cured tobacco leaf.

Those insects with immature stages that are dramatically different from the adults' have the opportunity of exploiting two or more different food resources during their lifetime. This diversity enhances their chance of survival, and it is no coincidence that most of the common insects such as beetles, butterflies, moths, flies, and wasps have a life history of this type. One of the most complex life histories of all is that of a beetle (Meloidae). The very mobile larva hatches from an egg laid at the roots of weeds and soon climbs a nearby plant stem in search of a flower head. Once perched on a flower, it patiently waits for the arrival of the right kind of bee and when the critical moment arrives, it stretches up and grabs any part of the bee's body that is within reach. Hanging on tightly, it is transported back to the bee's nest, whereupon it drops off and, without being molested, finds a cell where a young bee grub is being reared. The beetle larva enters the cell, kills the bee grub, and molts into a similar-looking legless maggot; it now has no need for legs, for the bee will feed it as its own until it is ready to molt into an adult beetle.

The social insects have been studied and admired for centuries. Bees, wasps, ants, and termites each have intricate societies in which different members are specialized for foraging, defense, and reproduction. The life of a worker honeybee is even separated into successive occupations; for example, during the first three weeks the young worker is confined to the hive, where it grooms the queen and her eggs, carries the eggs to the nursery, cleans out the hive, cools the hive by wing-fanning at the entrance, and attacks or walls-in intruders. Only after this apprenticeship is the graduate allowed to leave the hive and forage for nectar and pollen. The outgoing workers are informed of a good place to visit by an experienced returning scout who has located a rich source of flowers. While the species of flower is identified by scent on the body of the scout or by nectar regurgitated from its stomach, information on geographical location is imparted by a dance, which was first interpreted by an Austrian, Carl von Frisch, in 1923. If the source is really close to the hive, the scout dances in a circle, whereas if it is distant the dance takes the form of a figure eight. As the scout crosses its own tracks at the middle of the eight, it waggles its abdomen and emits a buzz. The rapidity of the dance and the number of waggles and buzzes tell the workers how far away the source is, while the direction of the eight on the vertical comb bears the same relationship to gravity as does the direction of the sun to the intended line of flight. So, when the naïve worker emerges from the hive, it already has information on the type of flowers it is seeking, in which direction it has to fly, and at what distance the flowers will be found. Other researchers have shown that the flight distance is measured not by air or land distances but by the expenditure of energy. As wind changes seriously upset their calculations, bees are rarely seen foraging on gusty days.

During the winter the temperature of the bees is regulated by how tightly they cling together; if the temperature rises, they separate to lose heat more rapidly, whereas if it drops, they form a smaller ball for conservation. When it is too cold to venture outside, their principal concern is to hold their excrement in order not to contaminate the hive. On the first warm day in spring there is a mad rush for the exit.

The ultimate advance in social organization is seen in the slave-master ants. A fertilized queen of *Polyergus* seeks out and invades the nest of *Formica* ants and with little apparent opposition kills the resident queen and ensconces herself in the royal chamber. The *Formica* workers immediately transfer their loyalty to their new matriarch and care for her eggs and young as meticulously as those of her predecessor. Of course, the only ants which now mature are the aggressive *Polyergus* adults which, lacking the ability to forage because of their large mandibles, continue to be cared for and fed by the *Formica* workers. However, it is not long before the efforts of the workers become weakened, for no new *Formica* workers are being raised now that their old queen is dead. At a critical level of labor shortage the *Polyergus* adults emerge in force and raid a neighboring *Formica* nest. The return home is always triumphant, for each *Polyergus* carries an immature *Formica* in its jaws to supplement their dwindling domestic labor force. Slave-masters indeed!

When you add to such behavior the fact that some ants use leaf fragments to act as spoons in which to carry soft food back to their nest, it is tempting to describe insects as "clever" or "intelligent" and begin to make comparisons between insect and human societies. We must beware of endowing other animals with human attributes, for such anthropomorphism is one of the cardinal sins of science, but, like some other sins, it is fun.

IV. An Insect by Any Other Name . . .

A little less than fifty miles north of Stockholm is the university town of Uppsala, where, in 1741, a professional doctor and amateur naturalist named Carl Linnaeus achieved his lifelong ambition to be appointed Professor of Botany. Linnaeus's interest encompassed all living organisms, and he had already achieved fame through the publication in 1735 of his *Systema Naturae*, which was a new classification of plants and animals. His town and country houses are still preserved as historic monuments and serve as a Mecca for science-oriented pilgrims from all over the world.

Though Linnaeus was aided in his work by student apostles, who were encouraged to travel to far-off lands in pursuit of new specimens, one can only be amazed at the enormous number of tropical plants and animals that were collected and named between 1750, when the voyages began, and 1777, when Linnaeus died at the age of seventy. For example, more than half of the thirty-eight species of snake now known from the island of Trinidad were named by Linnaeus during this period.

After Linnaeus's death and an extended domestic hassle, the greater part of his collection was sold by his widow to John Smith, an English naturalist who finally placed the specimens in the custody of the Linnaean Society in London, where they remain to this day. The balance of the collection, which contains substantial numbers of insects, is still at the University Museum in Uppsala. To examine a specimen pinned and handled by Linnaeus more than two hundred years ago generates a surge of emotion and awe—or is it just the fear of breaking it?

The most significant legacy left by Linnaeus was a system of naming each plant and animal with two names: a specific name which is peculiar to that kind of organism, and a generic name which is applied to all species considered to be closely related. He also organized animals and plants into the higher categories of families, classes, and orders, most of which are still in use. As was traditional for eighteenth-century scholars, his descriptions were all in Latin, an archaic practice which is still used by botanists today. Ideally, specific names should carry some descriptive connotation by which the animal can be characterized, for instance: *domestica*, the housefly; *damnosum* and *diabolicus*, which are two blood-sucking flies; and *irritans*, the human flea. Some persons would question whether *sapiens*, meaning wise, is a suitable name for man.

Carl Linnaeus was not the first naturalist to assign Latin names to organisms, but he was the first to do so within a comprehensive framework. Since then the naming of animals has become fraught with confusion. Naturalists have, in ignorance, described and named insects as new, though many have already been described and named. Other naturalists have assigned names that have been previously used for other animals. To describe an insect as new is relatively easy; it is knowing that it is really new that is difficult.

The problems with names became so pressing that in 1901 rules were formulated to regulate procedures and provide sensible guidelines for the future. It was decided that only names published after January 1, 1758, would be recognized, and, thereafter, the first occasion upon which a name was used would be accepted as the correct one. In general, this system has worked well, but every year or so an enthusiastic bookworm discovers that the widely accepted name of some common pest was first used for some obscure and little-known creature. In order to head off such potentially confusing changes it was decided in 1958 that any name that had been in common usage for fifty years could not be changed.

Of course, the naming of animals and plants has always been one of the more satisfying aspects of being a taxonomist, and even modern physicists are showing a touch of humor in their choice of such names as "quark," "charm," and "color" for their newly discovered subatomic particles. Perhaps the choosing of names allows the scientist to exercise a degree of innovation and artistry that is denied him in formulating the ponderous details of scientific description.

Toward the end of the nineteenth century it was fashionable among the wealthier Europeans to acquire a collection of butterflies, and the ambition of every rhopalocerist (butterfly collector) was to describe and name a new species or, at the very least, a new subspecies. It was in the pursuit of these goals that the well-heeled amateurs either paid scouts to collect for them in the fashion of Linnaeus's apostles, or purchased specimens from professional dealers and collectors. The commercial value of a specimen carrying a label, with the date and place of capture, has always been higher than that of one lacking such information, so, inevitably, many specimens had such data fabricated purely to enhance the price. Many of these capriciously labeled insects have found their way into our national museums. The problems they have created needn't be listed.

In their efforts to describe new species the butterfly collectors were reduced to examining each specimen scale by scale, the slightest deviation from normal being deemed worth the description of a new form or variety. Names proliferated, and inventiveness assumed new heights of ingenuity. Butterfly taxonomy has yet to recover from that excess of zeal. Feminine names abound, and one wonders if some of the relationships between the describers and the alleged goddesses, after whom their species were named, were less ethereal than was proper.

The International Code explicitly prohibits the use of any name that "when spoken suggests a bizarre, comical or otherwise objectionable meaning," and "no zoologist should propose a name that to his knowledge gives offense on any grounds." Was it, one wonders, the now suppressed generic name *Kismyas* that prompted that ruling?

So that later workers can be sure that names are being properly applied, the describer of a new species is required to deposit the actual specimens that he examines in a named repository. Types, as these unique specimens are called, have a very high scientific and monetary value, for they are absolutely irreplaceable. Museum administrators vie for the acquisition of types, for they feel that they add prestige to their collections. Charlatan taxonomists have been known to describe new species in the sole hope that a museum will bid a high price for the types. Fortunately, such happenings are rare and museums have very discriminating curators.

The full usefulness of types was brought home to a colleague whose lifelong study of fiddler crabs was published a year or so ago. One of the descriptions of a new fiddler crab, published by a nineteenth-century German zoologist, seemed most peculiar. The organism did not fit into any rational scheme of classification. When the chance came to visit the German museum which housed the type, all was explained. At some time prior to being examined by Herr Professor, the specimen had been dropped and severely damaged. Perhaps fearing retribution, someone had restored the specimen by the addition of parts from another crab. The subsequent description of the species as new is adequate testimony to the quality of the repair job. However, both the technician and the describer should have known better, for the repairer had glued a female abdomen onto a male carapace and claw. Many a young entomology professor has discovered the ease with which insect parts can be glued together, after having been duped by his students into embarrassing excitement over a "remarkable new insect."

Sometimes it is not easy to decide whether or not a mistake is a joke. For example, Linnaeus described a new butterfly from North America as *Gonepteryx eclipsis*. The specimen he examined is still in existence and has proved to be the common European brimstone butterfly with a hand-painted pattern on it. Did he know? Surely such an experienced naturalist would not have been duped so easily. Or was he? We will never know.

V. Insect Cemeteries

In the world of insects, geriatrics would be a poor investment, for old age is rarely their problem. Most insects meet their death by exposure to inclement weather, starvation, disease, or the jaws of the enemy. A very few are fortunate enough to join the ranks of the elite by being captured by an entomologist.

Because of their external armor, scarab beetles found in the tombs of ancient Egypt are better preserved than the mummies they accompany, and even without such ideally dry conditions many insect specimens have survived more than two hundred years on museum pins. The housing of preserved insects is a major responsibility of the museums of natural history, the larger mausoleums containing many millions of such corpses. Curators generally reckon that even the most common insect in the museum represents well over a dollar's worth of labor, for each specimen has been caught, killed, relaxed, pinned, spread so that its legs and wings are in the correct position, labeled, identified, and placed in the right drawer alongside its fellows. The less common insects are more valuable only on account of the difficulty of finding and catching them, for the other costs remain the same.

No curators take greater pride in their collections than the serious amateurs, some of whom have even been known to rent bank vaults to secure their treasures. One collection of small beetles I saw in Cambridge, England, was housed in thousands of European slide-tray matchboxes, all glued together to make floor-to-ceiling wall cabinets. Inside each drawer was from one to sixteen small beetles, each exquisitely positioned and glued to a rectangular card. Some of the insects were barely a millimeter long, yet every one of the tens of thousands of specimens had had its legs and antennae painstakingly arranged—a whole life's work.

The collecting urge can be powerful and may even exceed the boundaries of kleptomania. Some years ago I visited a well-known museum to look at its collection, and I had brought with me a most unusual insect, which I had been quite unable to identify. I took the specimen to an elderly, very revered entomologist and asked his opinion. He explained that as the specimen was so small, his examination would take a little while. I left my treasure with the expert and returned to my bench. It was not long before I felt an uneasy presence hovering at my side, and when I looked up, there was the rather flustered expert, apologetically explaining that he had just been transferring the insect from the vial to the stage of his microscope when it fell to the floor. Alas! The diligent search conducted by his staff of assistants had failed to recover it. He was most terribly sorry. I concealed my anger and disappointment and assured him that, as I was returning to Trinidad, I was sure that I could find another specimen (though I knew that I had found only one in ten years of collecting). Later that day at lunch I told my tale to some of the other scientists on the museum staff. They slapped their thighs with laughter. "Up to his old tricks again," they roared. "We can't wait for the old buffer to move on so we can see his collection." This was a new and totally unexpected facet of the real world from which I, as a naïve tropical entomologist, had been completely sheltered. I now realize that museums even have blacklists of visitors who are to be closely chaperoned from the time they arrive until the doors close safely behind them. Even in spite of their precautions, museums still suffer substantial losses at the hands of fanatical collectors, though by virtue of the nature of the specimens they lose, the curators can usually guess the identity of the thief. Patiently they then watch the obituary columns in the hope of making an eventual recovery.

The unexpected demise of amateur and professional entomologists has led to the loss of very valuable scientific material. Some collections containing type specimens have never been found, and the astute museum curator is always alert to the possible discovery of historic material. Such acute perception is typified by Dr. Harold Oldroyd's discovery of part of the Bracy Clark fly collection in 1962. Bracy Clark (1770-1860) was a professional veterinarian with an amateur interest in bot flies, whose larvae feed under the skins of farm livestock. Clark's descriptive papers were illustrated with colored drawings, but he neither deposited his specimens in a museum nor made adequate provision in his will for the bequest of his personal collection. So there were no actual insects with which modern material could be compared. By chance one evening at a Royal Entomological Society of London tea, Dr. Oldroyd met a gentleman who had just purchased an old insect cabinet from a local furniture shop. "Yes, there were some old-looking insects in the drawers, mostly beetles, but some flies, bot flies, I think." What a stroke of luck! Could it be the Bracy Clark collection? Subsequent comparison of the specimens with the illustrations showed beyond doubt that some of the missing insects had been found.

Much of one's time as a curator is spent in tracking down "lost" specimens, most of which are not really lost, but just mislaid or lacking distinguishing labels. Many of the early taxonomists failed to differentiate their types from their other specimens; hence, subsequent revisers have the problem of selecting the types from among the thousands of other similar-looking insects now held by the museums. The search involves guessing which museum has the material, knowing the locality and other data that should appear on the label, being able to recognize the handwriting of the author and the paper he selected for his labels, and sometimes even knowing what kind of pins he used. This kind of detective work is time-consuming but fascinating.

The hazard facing the museum specialist is that, in contrast to the general collector who learns less and less about more and more until he knows nothing about everything, the specialist learns more and more about less and less until he knows everything about nothing, at which point he risks seeming as lifeless as the insects he studies.

VI. The Collectors

Few professional entomologists seem to appreciate how supremely fortunate they are, for they are paid to do what most people regard as a hobby. Their relaxed life may explain why a recent United States study revealed that entomologists are among the longest-lived scientists. The secret of their longevity is, I believe, their almost imperceptible transition into retirement. Provided they retain their eyesight and access to a good library, entomologists can work right up to and even through their dotage.

The lives of insect collectors are never dull, for no matter where circumstances dictate they live, the insects are there too, waiting for them. In my youth I used to rush around the English countryside spending brief moments at places where insect rarities had been taken in years long past. I know better now. Charles Elton, the "Father of Ecology," showed that if you search hard and long enough in a reasonably diverse habitat you will discover an astonishingly large number of different animals. He and his students from Oxford found almost half the known fauna of England in an area about one mile square at Wytham Woods.

Rare insects are rarely rare; they are just hard to find. If any species were truly rare, how would the sexes locate each other? Surely they would have become extinct. Insects can certainly appear rare for such reasons as that they are adult for only a brief period of the year, that they are restricted to a very narrow vegetative zone, or that they are very well concealed from the human eye. I was forcibly confronted with the "look-at-home" rule while studying Schizopteridae in Trinidad. These petite bugs are hardly more than one-twenty-fifth of an inch long and were at that time almost unknown. A few of these tiny insects had been caught some twenty years previously on a forest floor, several miles from the university campus. So, faithfully, I visited the locality and returned with bags of fallen leaves, which

I then placed onto the open tops of large funnels. When I looked into the preservative below the funnels a few days later, eureka! There were the insects I sought. Elated, I returned to the forest, gathered more leaves, and collected more specimens. Then one day I noticed that there were several insects in the preservative under an empty funnel. After some thought I suddenly realized that the insects had been attracted to the electric-light bulbs and had flown in the open windows. It subsequently turned out that the insects lived in the roots of the grass on the rugby football field just outside the laboratory and were to be found over the whole campus. I should have looked at home first.

The tropics have been visited by numerous collectors during the past two hundred years, and certain localities have become famous for their apparent abundance of interesting insects. Tingo Maria, in Peru, is such a place, but the truth is that it is not the insects that are special, but the good hotel that North American and European collectors find so comfortable. As rich a fauna, and perhaps richer, awaits collectors in other vegetatively similar localities along the length of the tropical Andes.

Collectors are not so adventurous as our imaginations, for, like prospectors, they follow well-trodden and proven paths for fear of failure. The result has been that museums are full of insects from comparatively few tropical localities. That this is so can be verified by comparing the distribution maps of any pair of tropical organisms, for many of the spots on the maps will be seen to coincide. The greatest rewards await the explorer who strikes out into virgin territory where collectors have not ventured before. Such novelty is becoming easier as new roads open up such regions as the Amazon Basin.

However, in many parts of the world, travel is handicapped by the dangers of assassination, kidnapping, or political confinement. In 1963 I was in Colombia collecting butterflies in the valley of the Magdalena River. The day after I returned to Bogotá, the banditos rounded up a busload of local people from the village where I had stayed and executed the men. The only man who was spared had briefly escaped detection by hiding under the voluminous skirts of a very substantial lady. Upon discovery, he was deemed so lacking in manhood that he failed to qualify for the routine emasculation administered to his fellows. I cannot believe that I would have shown such extraordinary presence of mind.

Later, on the same trip, I was jailed by the Ecuadorian army because they thought my butterfly work was merely a cover for espionage. Once convinced of my innocence, the soldiers were anxious to help, and I ingratiatingly accepted the mangled specimens they had caught with their steel helmets. Poor material for genetics research, but excellent for international and personal relations.

As people, insect collectors are intensely interested in their work and certainly qualify for the "crazy bug hunter" label of skeptics. But they are to be envied, for they find a wealth of interest and satisfaction wherever they happen to be. Their principal fault is, perhaps, that they take their calling too seriously. One famous English entomologist used to employ young male scouts, whose main responsibility was to search for interesting insects. When an insect was located, one boy would secure it with an inverted vial while the other ran to fetch their employer, who could then place the cap on the vial and truthfully write "Taken by [his own name]" on the data label. At the other extreme is the tropical visitor who sits in his rocking chair sipping his nightly dose of quinine and gin, periodically prompting his wife with "Isn't it time you checked the lights, dear?"

When I was working on the insect pests of cotton in Nigeria, I used to play cricket almost every Sunday during the dry season, which in Zaria lasts about five months. Being more enthusiastic than skillful, I was usually placed somewhere in the field where the captain thought the ball would be most unlikely to be hit. During these prolonged spells of inactivity I used to "fish" for tiger-beetle larvae. Tiger beetles develop from grubs which live in a pencil-wide vertical shaft in the ground, about fifteen inches deep. During the night or the cooler parts of the day the grub blocks the opening to the shaft with its head and grabs any unsuspecting insect that happens to be trotting by. At the approach of human footsteps, it retreats to the bottom of its tunnel and hides in a side chamber. The fishing game consists of poking a long grass stem down the shaft and waiting for the grub to bite it, whereupon a "fisherman's strike" will sometimes haul up the struggling larva by its teeth. Provided the captain does not change the placing of his fieldsmen, the growth rates of several grubs can be monitored during the cricket season and the away-from-home games provide an opportunity to sample more remote populations.

Caricatures of entomologists traditionally show bespectacled old gentlemen wielding butterfly nets, and, indeed, a net is the most useful single instrument. The best insect nets are made of fine diaphanous nylon in order to present the smallest resistance and weight for aerial maneuverability. Fancy nets have telescopic handles that extend up to fifteen feet, though at that height they are very unwieldy. Nets for sweeping insects from rank vegetation have canvas bags with edges bound in abrasion-resistant leather. A neat device for snapping up insects on flower heads and leaves was invented by Michael Samways. It consists of a pair of opposed tea strainers attached to the tips of a pair of fire tongs. But the ultimate in indolence is the use of the "Autokatcher," an insect net that fits on the top of a car so you can collect as you drive.

During recent years the use of suction traps in which an electric fan draws air and insects into a killing jar has proved very useful, because all small, flying insects are caught indiscriminately. Only the more powerful fliers can sometimes avoid the draft. The problem with the traditional black-and-white light traps is that the catch clearly depends on whether or not the insects find the light attractive. The two sexes of a single species sometimes show dramatic differences in their attraction to lights, which leads to false conclusions about their natural abundance.

Both suction and light traps can be constructed so the catch is collected separately during different hours of the day or night. Such devices can reveal interesting information about the flight habits of the insects. For instance, while collecting in Trinidad with a simple homemade light trap and timer, the female bark beetles thought to be responsible for the transmission of witches'-broom disease of cacao were caught only during a twenty-minute period at dusk and at dawn. With such a short flight time, trapping the insects with lights became an economic possibility.

The catch from lights hoisted to different heights in the forest canopy shows that some insects are strictly zoned. In East Africa massive steel towers have been built rising up to 125 feet into the trees so that mosquitoes carrying infectious diseases can be sampled. In Trinidad the highest light I could install was about 50 feet, because that was as high as I could accurately throw a mango seed with a thread tied around it. The results of the catch showed that a collector with a trap only at eye level would have overlooked at least twenty-five per cent of the insect species present.

A slick-sided vessel, buried in the ground and baited with fresh feces, rotting banana, or a mixture of stale beer and treacle, will catch a remarkable selection of insects. Treacle and beer (or blended Jamaican rum) is a traditional recipe with moth collectors, who paint the brew on tree trunks and fence posts early in the evening and then, after dark, visit the sites with a flashlight. Drunken insects, particularly ants and moths, can then be picked off with the fingers. The brilliant metallic-blue *Morpho* butterflies of Brazil and Central America are notoriously difficult to catch on the wing, yet they are so partial to a beer-and-banana mash that they, too, can be hand-collected.

Sooner or later, even if only briefly, every collector wonders if he or she is contributing to the extinction of some insect species. While it is true that the phrase "endangered species" used to conjure up the image of exotic birds being hunted for their plumage and rhinos being carved up for their horn, we now have endangered species of plants, fish, amphibians, insects, and, in fact, members of almost every group. Not only the more spectacular insects are endangered, for the 1973 list of British Rare and Endangered Species contains dragonflies, grasshoppers, and beetles, as well as butterflies and moths.

Fourteen of the sixty-eight resident species of British butterfly are endangered. Did Simon Barnes and I contribute to this disaster some thirty years ago? Probably not, for the really hostile pressures on these insects are not from collectors but from environmental damage by industrialization and human growth. Only when a particular insect is hunted intensively for commercial trade is collecting alone likely to endanger it. Such seriously threatened insects are the beautiful birdwing butterflies of the East Indies which are now protected by law. Distressingly, the act of protection makes these gorgeous animals even more valuable and vulnerable to poaching, as is attested by the fact that they are still prominently displayed in the naturalist shops of North America and Europe.

Though we know so much about insects, there are simple questions to which we have no satisfactory answers. Why are some aphids attracted to yellow paint? Why are some insects attracted to light while others are not? Why is a moonless night with a little rain just before dusk the best time for trapping insects with lights? Why are some butterflies attracted to dried heliotrope plants? Perhaps questions such as these are best left for the research scientist, but thousands more are within the solving power of the naturalist. One does not have to be a collector to study insects, but it certainly helps, because it provides the motivation to spend the time searching and waiting.

There are probably more than a million different species of insect, neither endangered nor rare, awaiting our attention. Some of the most famous and productive entomologists have been clergymen, engineers, clerks, auctioneers, accountants, and businessmen. All that is needed to enjoy the excitement of entomology is the time and the initiative. R. H. L. Disney found both and published a paper, in the *Entomologist's Monthly Magazine*, entitled "Flies Associated with Dog Dung in an English City." Now there is a challenge; surely you can be more innovative than that!

4

5

6

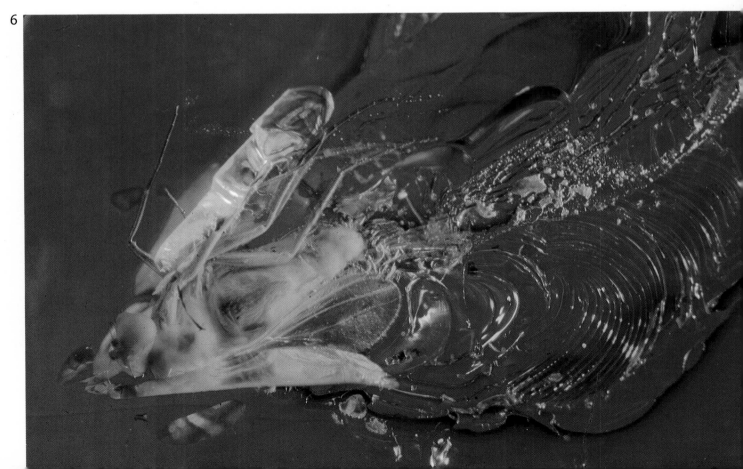

7

8 9

10   11

13

14

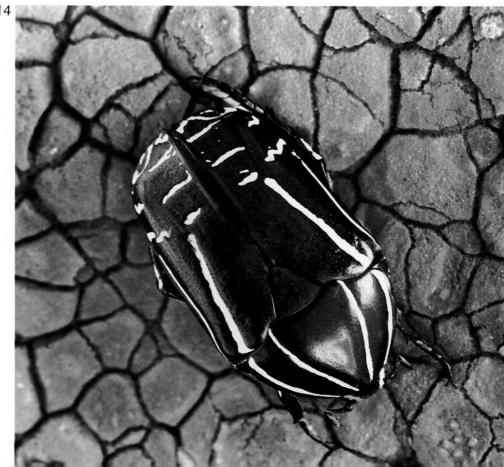

18

19

20

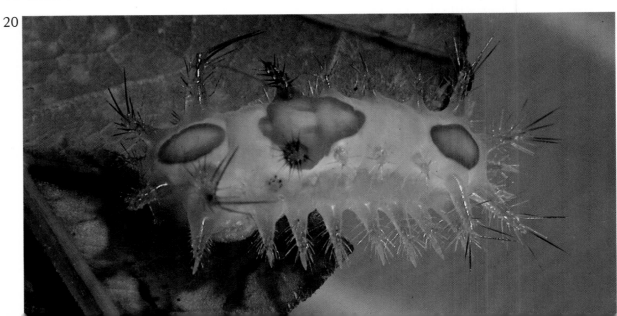

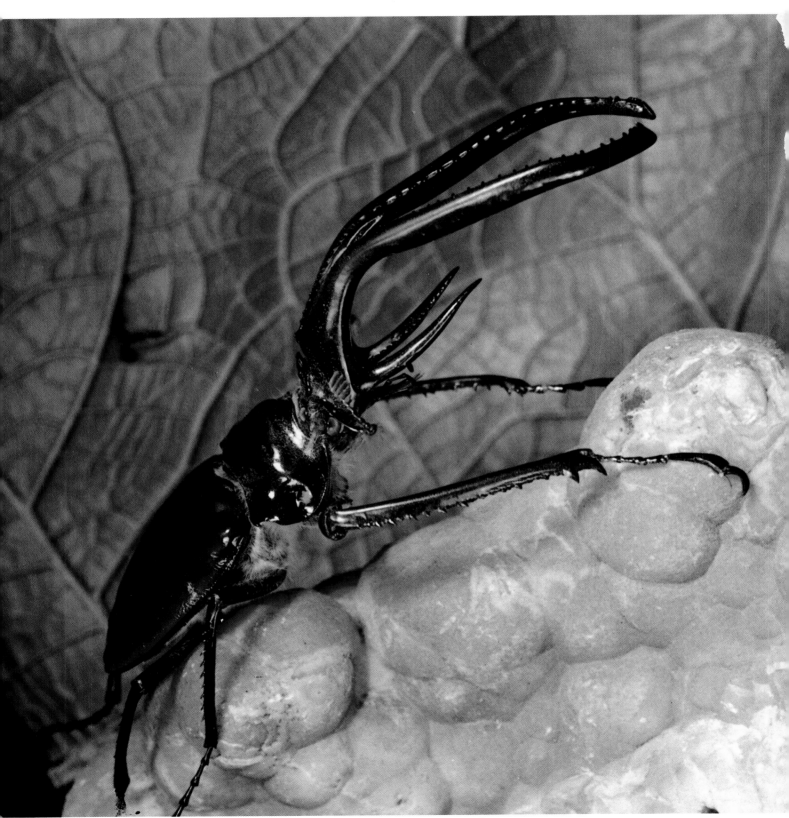

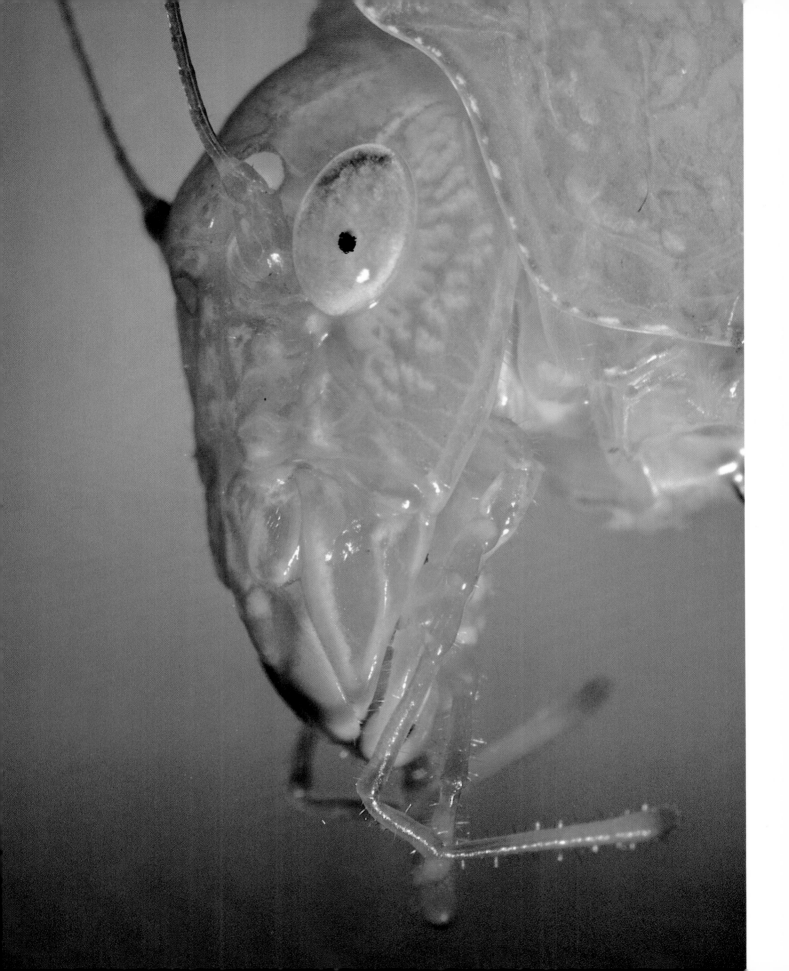

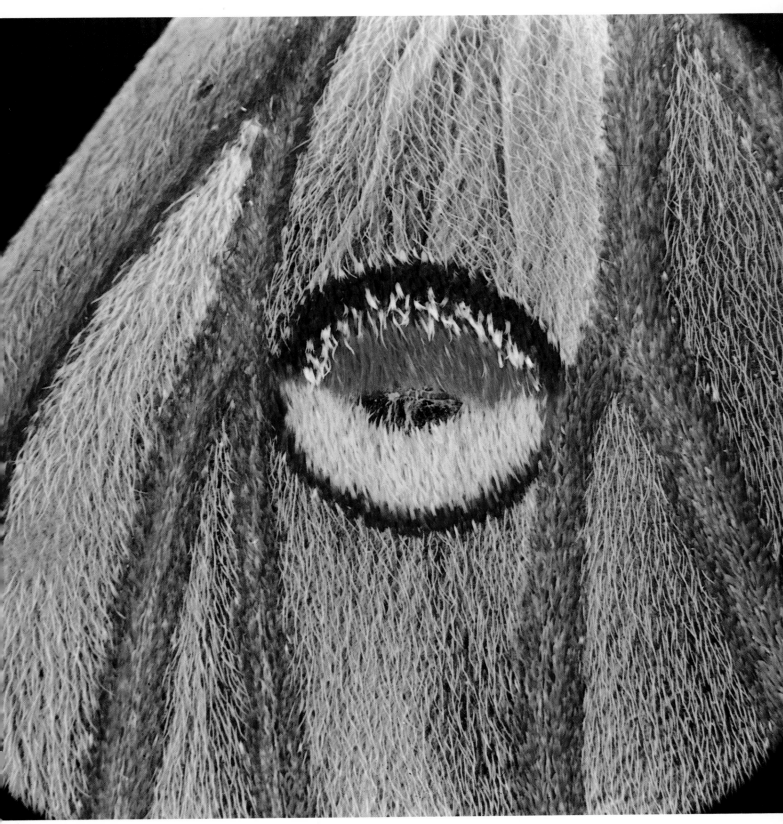

37

41

42

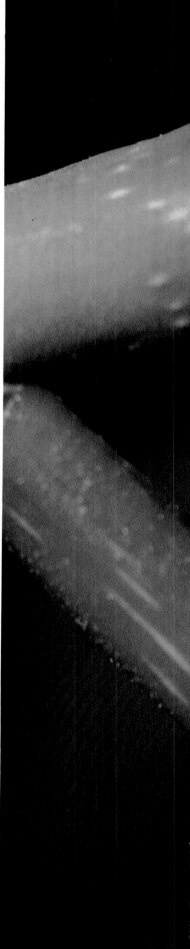

43 44

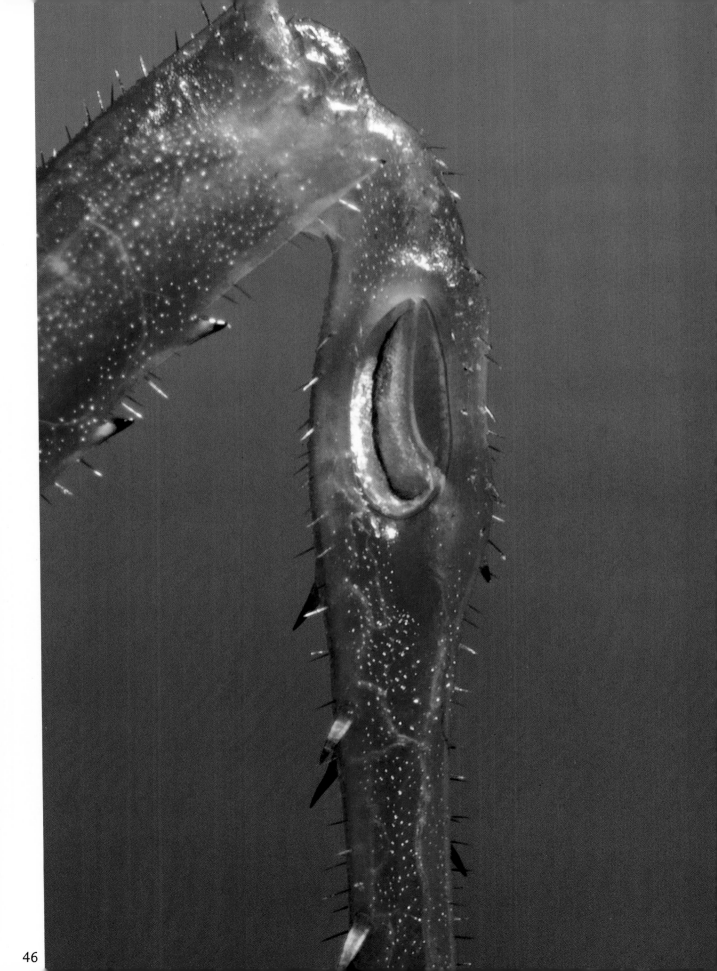

49

50 51

53

54

55

62

64

66

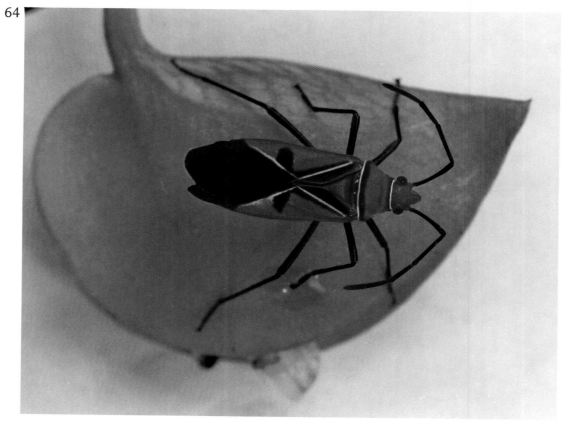

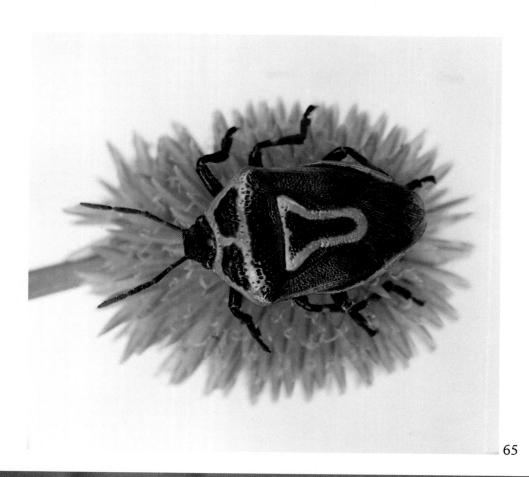

63

65

68

67   69

70

71

74

75

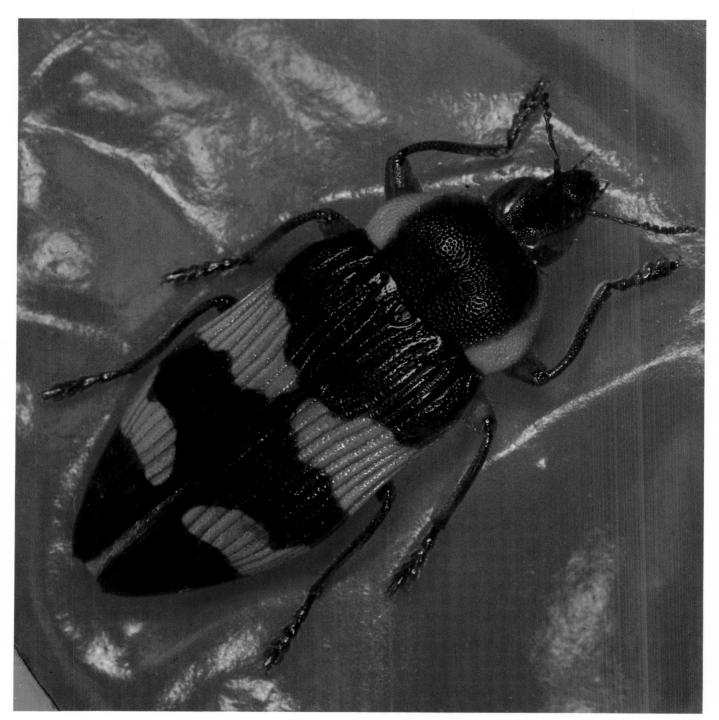

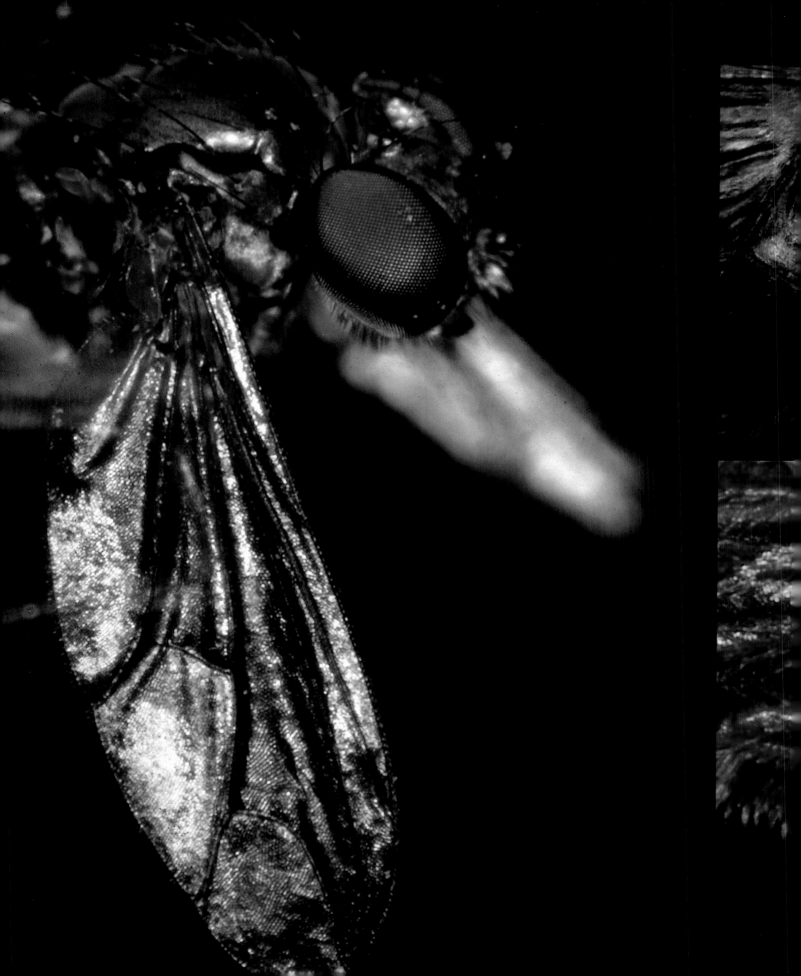

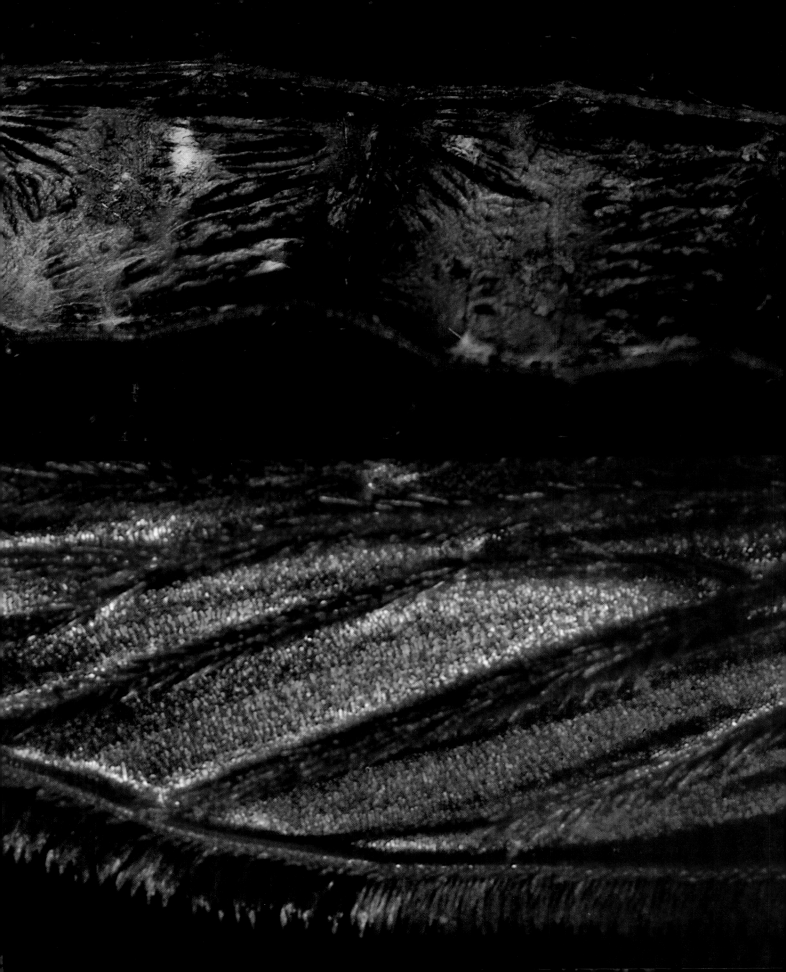

Notes to the Plates

1. In this leaf's-eye view of *Leptocoris trivittatus*, the sucking tube with which the bug draws up plant juices can be seen projecting from beneath the front of the head. Though these insects puncture leaves, their principal nourishment is drawn from the seeds of the box elder, either fallen or while still on the tree. The red pigment in the insect's skin is obtained directly from the box elder. If the insects' numbers become excessive, they disperse and sometimes spread to fruit trees, where they are considered minor pests. Conspicuous insects are often evil-smelling, but these are quite odorless. As fall approaches, the gregarious adults are common house guests as they look for a cozy place in which to hibernate. Though harmless, when they invade dwellings in large numbers they are a great nuisance.

2. "Little fleas have lesser fleas, upon their backs to bite 'em!" So says the song, but in this case they are mites biting the leg of the Bolivian bee *Euglossa piliventris*. Mites are common parasites of the larger insects. However, not all the small animals found living on the cuticle of insects are actually feeding; some are just stealing a ride.

3. This *Campososternus gemma* from China is one of the many dazzling metallic members of the family Elateridae, which get their nicknames, "click beetles" or "skipjacks," from the clearly audible noise they make when righting themselves after being turned on their backs. Using a spine and groove on the lower surface of the body, an upside-down beetle can snap its body so explosively that it is projected six inches into the air. If it lands on its feet, it makes good its escape, but if not, it tries again and again and again.

Pyrophorus is a large South American click beetle, about an inch and a half long, which has three luminous spots, two on top at the front and one underneath at the back. At night the light emitted by them is so bright that one can read a newspaper while sitting near a glass jar full of beetles, and it is alleged that during the building of the Panama Canal a surgical operation was conducted by beetle light when all other sources had failed.

4. Crazy paving? No, just a microscopic view of the wing of the grasshopper *Lophacris gloriosa* from Panama (see plate 66). These fine crossveins help to strengthen the wing for flight and also create the protective illusion of leaf venation when the insect is at rest.

5 & 6. About thirty-eight million years ago both these insects were trapped in resinous gum exuding from pine trees in northern Europe. Now embedded in amber, which is fossilized resin, these specimens prove to us that insects essentially similar to modern forms have existed on earth for a long, long time.

Flies are the insects most frequently found preserved, but other types were also trapped. The turbulence around the corpse in plate 6 re-creates the death throes of the drowning insect.

7. The hind legs of this stingless bee (Meliponini) from Costa Rica can be seen to be enlarged for carrying pollen. Different species of these bees have colonies varying in size from less than thirty to more than eighty thousand individuals, and their nests may weigh nearly one hundred and fifty pounds. Each nest, whether it is constructed in a hollow tree, in a crotch of the trunk, or in a hole in the ground, usually has an entrance/exit tunnel, as in the illustration. The tunnels, though normally only an inch or so long, can, in exceptional cases, be as long as fifteen inches. The tunnel walls are made of wax, resin, and earth. Before post-Columbian settlers introduced the European honeybee, Indians of Central America and southern South America domesticated the stingless bees as a source of honey and wax. Though they possess only a vestigial sting, when handled these bees can exude a very irritating fluid and bite very hard.

8. This wasp is a typical member of the Ichneumonidae, which lay their eggs in the bodies of a wide variety of other insects. As the parasitized host always dies, the attack is really a prolonged predation. The term "parasitoid" has been coined to describe this lifestyle. The host survives for quite a long time while the young wasps mature inside it, for the maggots take care to avoid the vital organs until they are nearly ready to emerge. Many a small boy has been disappointed to find that his laboriously nurtured caterpillar yields only a dull-looking wasp instead of a resplendent butterfly or moth. The female wasps have an uncanny way of knowing which caterpillars have already been parasitized, and on these they waste no eggs.

9. This orchid, a species of *Dendrobium*, is a native of the Sunda Islands in the East Indies, but in Hawaii, where the orchid has been introduced as an ornamental, the naïve local bees do not realize their limitations and habitually become stuck in the flower entrances. Each flower on this inflorescence has a bee sentenced to be "hanged by the thorax until dead" for being so bold as to think it could reach in and steal the nectar with impunity. In Sunda these flowers are pollinated by smaller native bees which, of course, do not become stuck and are then able to pay for their nectar reward by transferring pollen to other flowers.

10a & b. These two frames from a motion picture show a Panamanian *Protopolybia pumila* wasp bailing out water from its nest. The nests are usually very carefully designed and constructed so that rain cannot enter, but, in exceptional cases, leaks do occur which require remedial activity. *Polybia* wasps are famous for their ferocious disposition and painful sting, but there is recent evidence that the sweat of some lucky people pacifies the wasps and renders the humans immune from attack.

11. The botanic gardens in Singapore, where these bees are visiting the flowers of the water lily *(Nymphaea lotus)*, are among the oldest in the world. Early British colonials established botanic gardens all over the tropics, and these botanic gardens still exist as oases of biological culture.

Water-lily flowers are open for relatively short periods of time, so the time of their opening has to coincide with the time of flight of their insect pollinators. Some bees have regular daily routines and travel predictable pathways at the same time each day so that they arrive at flowers shortly after the blossoms open. Ecologists call this patrolling of regular pathways by bees traplining.

12. Members of the subfamily Cassinae (Chrysomelidae) are usually called tortoise beetles because of their shape and heavy armor, but this *Tauroma casta* from Costa Rica is unusual in that it has a pair of finlike projections from the wing covers. The females of some species of cassid beetle are interesting in that they show elementary maternal care by guarding their eggs and young and by attempting to fend off enemies. The larvae of other species have the unusual habit of camouflaging their bodies by covering them with their own feces, an optical confusion which is enhanced by the addition of their own cast-off skins. In extreme cases the larva may be totally concealed under this pile of refuse.

13. Leaf beetles (Chrysomelidae) come in an extraordinary number of shapes and sizes. This turtle-like *Stolas imperialis* is Brazilian. It has been said that when God made the insects He was inordinately fond of beetles, or at least it would seem so, for there are nearly 300,000 known species, which is more than a third of all recorded insects. There are more kinds of beetles than all animals that are not insects added together. Beetles are also the largest and smallest insects: Goliath beetles are larger than some small mammals, such as mice, while the Ptiliidae are smaller than some single-celled animals.

14. Driving a pin through the armor of *Rhabdotis semipunctata* from South Africa is usually more than the metal—and human patience—can bear. These beetles are often found feeding on flower heads, where their bright color and large size make them very conspicuous. Presumably, their heavy armor renders them virtually indestructible by predators.

15. While the back of most treehoppers (Membracidae) is drawn out into hooks, spines, and processes, the pronotum of *Oeda inflata* from Tingo Maria in Peru expands into a large leaflike covering. The specific name for this treehopper is aptly chosen, but so is its common name—balloon treehopper. The reticulate design on the surface is certainly similar in texture to that of a leaf, so in this case it is easy to speculate that the structure is purely protective. The case of some other treehoppers is not so easily explained (see plates 16, 33, 35, and 67).

16. Pre-Columbian art? Not really, but against the right background this Mexican *Sponogophorus hoffmanni* is all but invisible. Is the filigree over-done? Perhaps, but most of the Membracidae are so extensively or-namented that they invite that question a thousand times. Known as treehoppers, the adults feed on the sap of trees, but during immature stages many species feed on the rank vegetation on the ground.

17. As this colony of polybine wasps grows in size, the tiers of cells will almost fill the kilnlike envelope (cut away in this view to expose the internal architecture). Presumably, it is because *Polybia* wasps are ubiquitously feared in South America for their agonizingly painful stings that the Yukpa Indians have chosen to use them in a symbolic ritual: the father of a newborn boy must provide wasp larvae for the ceremony, which necessi-tates the collection of several nests of *Polybia*, during which no aids such as fire or smoke may be used. The stoicism and silent tolerance of pain exhibited by the father is thought to impart desirable qualities to his infant.

18. Because of their covering of long hairs, the caterpillars of many temperate tiger moths are known as woolly bears. The hairs protect the caterpillars from such predators as birds, for they are very sharp-tipped and easily dislodged, as many a young collector has discovered to his discomfort. The tiger-moth family, Arctiidae, is notable for the beautifully colored adults and caterpillars, such as this *Seirarctia echo*. In the tropics there are arctiid adults with a dimpled plate on each side of the body, which, when popped in and out by the contraction of a muscle, emits alternately ascending and descending "zips." If the moth is in flight when it hears the echo-locating sounds of an approaching bat, it zips the plate in and out, closes its wings, and drops to the ground. There is some controversy over whether the sound confuses the bat by jamming its locating system or is a warning that the moth is bad-tasting. The drop to the ground may be an extra precautionary measure in case the bat is naïve and has not yet understood the message.

19. The inchworm, or earth measurer, is a childhood charmer as it loops its way along, alternately bringing its four hind legs up to its six front ones and then extending the front ones forward. These caterpillars mature into rather dull-looking moths of the family Geometridae, though some are green, a very unusual color among moths and butterflies. The green colors are liable to fade in collections and invariably turn yellow if the insects are killed with the traditional cyanide killing jar.

During the day, when not looping or feeding, these caterpillars remain motionless, frequently clasping a branch with their hind legs and projecting out at an angle, just like a broken twig. If an insectivorous bird finds one of these protectively colored mimics, it will spend the rest of the day inspecting every twig of similar size in the often vain hope of finding another inchworm.

20. The caterpillars of Limacodidae (Eucleidae) are known as slug caterpillars because of their smooth fleshy bodies, but no slug has such poisonous spines. The hollow spines have venom glands at their base, and if they are handled very gently they have no effect because the tips are intact, but if they are grasped firmly, the collector's hand will be poisoned and will swell up like a cow's udder, with the fingers projecting like teats. The pain is excruciating. A number of these widely distributed moths have caterpillars resembling this saddleback from West Africa.

21. The efficient work of a leaf skeletonizer leaves only the hard indigestible veins. The caterpillars of several families of small moths feed in this way, most notably, perhaps, the grape-leaf skeletonizer *Harrisina americana*, which is a commercial pest. There is a quiet beauty in their finished product, but it does not match the perfection achieved by microbial decomposers such as fungi and bacteria, which produce the elegant fall leaf patterns.

22 & 23. The Lucanidae has many members, such as *Chiasognathus granti* from Chile (22) and *Cyclommatus imperator* (23) from New Guinea, with the mandibles developed into the huge hornlike structures seen here. Little is known about the function of these jaws except that they are usually much larger in the male. In some species the males fight each other over a female, the vanquished male being carried away from the female in the jaws of the victor. The bizarre mandibles and other projections make these beetles particularly clumsy, and one wonders how they have managed to survive.

The occurrence of *Chiasognathus* in southern South America, South Africa, and Australia had puzzled zoologists for years, but their existence in all three areas makes sense, since we have realized that some 200 million years ago these three continents, together with India and Antarctica, were united over what is now the South Pole.

24. The microscopic surface sculpturing of the transparent wing of a fruit fly *(Drosophila)* reflects a rainbow of colors when illuminated by electronic flashlight. The texture of the wing surface gives the illusion of being covered with fine scales, but each highlight is no more than a minute hillock on the wing surface.

Fruit flies are best known for their involvement in genetic research and as a nuisance on overripe fruit. More scientific papers have been published on fruit flies than on any other type of insect. In fact, some years ago Paul Ehrlich, of contemporary environmental fame, suggested (tongue in cheek) that attempts to discover more species of insects should cease, for if we were to study the already known species with the intensity that fruit flies have been researched, the library shelves needed to hold the scientific papers would cover the whole globe.

25. This view of a Peruvian arctiid moth reveals the exquisite color vested in the individual scales with which the wings are clothed. Pale blue coloration in butterflies and moths is usually produced by pigments which fade with the passage of time, but the brilliant hues are due to optical effects in the structure of the scales and will last the life of the specimen. It is only insects with these permanent structural colors that are used in the jewelry trade.

26. New Guinea is a paradise for tropical insect collectors and holds many secrets still to be discovered. Here we see an adult and juvenile member of the principally tropical family Flatidae. Like the young of many nearly related families, these juveniles secrete numerous filaments of white wax which bestow on them an ethereal quality. Presumably, the wax acts as a protective covering that is distasteful to birds and other predators, and serves as a measure of concealment in which the insect's outline is camouflaged.

27. This top view of the head of a bluet damselfly (Coenagrioniidae) shows the great development of the eyes at the extremities of the transverse head, an arrangement which provides excellent stereoscopic vision for judging distance. A slight movement of the head gives a full 360-degree field of view. Damselflies, and their larger cousins the dragonflies, are predators which catch their prey in flight. They need excellent vision, and these eyes supply it. The small size of the antennae in the front of the head suggests that the sense of smell is not very well developed in these sight-oriented insects.

28. The eyes of this deerfly, from Washington, D.C., show the complexity of the color patterns produced by the surface sculpturing. It would be convenient for purposes of identification if these patterns persisted after death, but the drying of the cuticle causes changes in the surface and the colors soon disappear. In most parts of the world, but notably in Africa, these insects transmit serious diseases to man and his domesticated livestock.

However, with or without disease, a bite from a tabanid is a bite to remember. In parts of Russia these flies can be so numerous that farm workers are forced to till the soil at night, at which time the insects are inactive.

The larvae of Tabanidae, which includes this species of *Chrysops*, are restricted to damp situations, so the adults are most frequently encountered near freshwater ponds, streams, and marshes. They appear to fly extremely fast, but in reality they rarely attain speeds of more than twenty miles an hour. In 1926 it was claimed that a fly was timed speeding at 820 miles an hour. The claim was discredited, but for many years it was an often quoted "scientific fact."

29. The compound eyes of insects contain up to 30,000 individual lenses, each of which forms an independent image. In 1891 Siegmund Exner photographed a cross through a church window and his colleague, Sir Edward Poulton, using a glowworm's retina as his lens. Both upright images—the cross and the colleague—were clearly recognizable and composed of multiple contributions to the whole picture, somewhat after the fashion of a halftone photograph in a newspaper. It is likely that at least some insects can see their surroundings very well. When we look at the eye of a live insect, one whose compound eyes contain a relatively small number of lenses, those lenses looking directly back at us reflect no light and appear jet black. It is this optical effect which produces the false "pupil" of the katydid *Microcentrum rhombifolium*, pictured here. As the viewer moves around the eye, the pseudopupil follows.

30. The horizontal line across the eyes of this horsefly is, as in plate 28, an optical effect produced by the surface sculpturing of the lenses. Several species of *Tabanus*, including this specimen from Washington, D.C., have intensely green eyes that occupy the greater part of their heads. It is not surprising that these flies should be commonly known as greenheads. Females must have good vision, for they locate their blood-meal victims by detecting their movements; however, the eyes of the males are larger, even though they feed only upon nectar, plant-wound sap, and aphid honeydew. The reason for this inconsistency is obscure. Also mysterious is the fact that the males of many species of Tabanidae are so rare that in some cases they are unknown to man. Presumably the female tabanids can find them without difficulty.

31. *Laternaria phosphorea* from Peru, as well as similar species from other parts of South America, have puzzled entomologists for more than a century. It was once thought that the grotesque heads of these insects, known as peanut bugs, alligator bugs, and lantern flies, were luminous. Peter Parley's *Tales of Animals*, published in 1839, claims that "this beautiful insect is a native of Surinam and many other parts of South America, and during the night diffuses so strong a phosphoric splendor from its head or lantern, that it may be employed for the purpose of a candle or torch; and it is said that three or four of these insects tied to the top of a stick, are

112

frequently used by travellers for that purpose. A single one gives light enough to enable a person to read." However, no contemporary observations have confirmed these claims, and even if the heads of these insects are sometimes truly luminous, the source of the light is most probably from bacteria living on the surface of the skin or in the intestines.

No reasonable suggestion has yet been offered for their strange shape. It is indeed true that the head does look like a peanut or alligator, but it is inconceivable that any insect predator could be deceived by such an illusion. Who ever saw a peanut or an alligator resting on a tropical tree trunk, or on a *Heliconia* inflorescence as in this picture? We can speculate that the eyespots on the hind wings, which are revealed during the display as shown in the picture, might startle an overinquisitive bird, but the insect is, at most, only three inches long.

32. Eyespots are common among butterflies and moths and may even come complete with "pupil," "iris," and "eyelashes." It is unusual to find the veins picked out in a different color as they are in this moth. The veins serve as strengthening girders to give rigidity to an otherwise flexible membranous wing. Blood flows through the veins only just after emergence from the chrysalis, so that the term "vein" is rather misleading. The distribution of this *Graellsia isabellae* is restricted; it is known only in the Spanish and French Alps.

33. Treehoppers are bizarre insects; their backs exhibit fanciful shapes, such as spines, hooks, and horns (see also plates 15, 16, 35, and 67). Theorists have attempted to understand the function of these outgrowths: some say that they serve to camouflage the insect; others speculate that special sense organs are embedded in them. If the latter theory is indeed true, we have to confess that we do not know what it is that they are sensing. This immature *Alchisme grossa* from Costa Rica is additionally interesting because its four spines, projecting from the wing-bearing segments of the thorax, will not persist when the immature insect molts into an adult.

34. Resembling a World War II warship, this immature South American grasshopper is conspicuous when it sits on a purple flower, but it is virtually invisible when it is motionless in grass. Notice the false "pupil" in the eye—an optical effect due to the construction of the insect's compound eye (see also plate 29). This little grasshopper is less than half an inch long, but some South American species attain a length of four inches when they mature and are so powerful that, if grasped only gently, a kick with their spiny hind legs can cut the palm of a hand wide open.

35. The extraordinary development of the pronotum in treehoppers is again displayed in this *Emphusis bakeri* from Mindanao, which is reminiscent of a sign of the zodiac.

36. Protective coloration alone is not enough, for insects must adopt appropriate attitudes to complete the illusion of absence. This small moth from New Guinea can be seen here clearly, but its abdomen is elevated in such a way that the normal outline of the insect is lost. On a tree trunk or twig it would blend beautifully into the background and perhaps escape the notice of a foraging bird. Very little is known of the biology of most tropical moths, especially the more inconspicuous ones such as this.

37. *Fulcidax bacca*, a chrysomelid beetle from Brazil, looks more like a Christmas-tree ornament than an insect. However, in its sun-flecked natural setting it is probably all but invisible.

38. Diamond beetles are weevils with iridescent metallic scales over their bodies. Spectacular examples are known to come from Australia, the Orient, and South America; this specimen is from Samar in the Philippine Islands. The family of weevils, the Curculionidae, contains more than 40,000 species and is the largest known family of insects. No single person can hope to become a comprehensive weevil expert in a whole lifetime of study, so entomologists interested in the group specialize in a small number of species. The biology of only a very few tropical weevils is known.

39. Each time a female lacewing (Chrysopidae) is about to lay an egg, she first touches the leaf with her abdomen and extrudes a drop of a substance that hardens rapidly when exposed to air; she then elevates her abdomen and draws the glutinous material out into a thread, at the top of which she lays her egg. High above the surface of the leaf, her eggs are thus protected from the casual wanderings of predators. The photograph shows a young aphid (left) and (right) a dead adult. Aphids form the staple diet of both young and adult lacewings.

40. A caterpillar's lot is not a happy one if it should be parasitized by *Apanteles* (Braconidae). Some weeks previously, this hawkmoth caterpillar (Sphingidae) was approached by a small wasp which laid one or more eggs inside the caterpillar's body (some braconid wasps lay single eggs from which emerge twenty or thirty maggots, a procedure known as polyembryony). With the maturation of the parasites (see plate 8 caption) the maggots emerge and spin white silken cocoons on the outside of the now dying host caterpillar. In a few days an adult wasp will emerge from each cocoon, mate, and begin the cycle anew.

41. One of the many beautiful lacebugs of South America, this species of *Leptodictya* is commonly found on grasses and bamboos. The exquisite sculpturing of the cuticle is quite breathtaking when seen under the microscope. The late Carl Drake made a lifetime study of lacebugs, and his many scientific papers are enhanced by superb pen-and-ink sketches of these tiny insects, most of which are less than one-third of an inch long.

42. Is this Peruvian mantis preying or praying? In the eighteenth century Carl Linnaeus named the European mantis "*Mantis religiosa*," but either word would do, for all stages of development are predatory on other small animals. The females of some species carry their eating habits to excess, for during the act of mating they devour their male partners, head first. This unusual habit is not without its reward, for nerve centers which prevent movements of the male abdomen are located in the front part of the body, so as the meal progresses, the excitement increases—but only for the chauvinistic female!

43. These worker ants *(Oecophylla smaragdina)* are beginning to make a nest of leaves on Magnetic Island off the coast of Queensland, Australia. While the workers press the edges of the leaves close together, several ants, each holding an ant grub in its jaws, walk a zigzag path from one leaf to the other, touching each leaf in turn with the sticky silk extruded from the mouths of the grubs. In this way the leaves are sewed together into a compact and strong dwelling for the colony. The caterpillars of moths are occasionally found living inside some nests, where they feed on the immature ants. Strangely, they are tolerated, perhaps because they spin a strong web of silk over the inside of the nest and so strengthen it and render it more waterproof.

44. Figs were being cultivated in the eastern Mediterranean some six thousand years ago, yet their dependence on their relationship with insects is one of the most complicated known in domesticated plants. The popular Smyrna fig, which is now grown commercially in Turkey, Greece, Italy, and California, depends for its fruit on *Blastophaga psenes*, a minute wasp about one-twenty-fifth of an inch long. The laying of an egg in a specially adapted sterile flower of the wild caprifig promotes the formation of a gall in which the wasp grub develops to adulthood. Males are wingless and emerge from their nursery to live only long enough to mate with neighboring newly emerged females. The females are winged and become dusted with caprifig pollen as they leave to search for a suitable site in which to lay their eggs. If they should erroneously visit a Smyrna fig, which has no special gall-forming flowers, pollen is transferred and fruit is set. To enhance the chances of successful pollination, most fig growers hang branches of the wild caprifig among the foliage of the Smyrna trees so the emerging females are almost certain to visit the flowers of the commercial crop.

It would be tempting to identify the two wasps on the fig in the illustration as *Blastophaga*, but they are not. The tiny wasp on the left is a parasite of the immature stages of other insects, whereas the other wasp is one of a group which contains both parasites, parasites of parasites, and gall makers. Some of these minute insects even lay their eggs in the eggs of such other insects as butterflies and mantids.

45. "Hello! Anybody home?" Though it belongs to the family which contains the feared locusts of Africa, this species of *Acrida* is more interesting for its strange head shape. Almost all grasshoppers feed on vegetation, and those which form destructive migratory swarms are called locusts, but even those which do not swarm can become so abundant that they, too, can be serious pests.

46. Differences in the time it takes for various sounds to arrive at the "ears" on the front legs of male and female katydids allow the insects to estimate the direction from which the sounds are coming. Only the males sing songs which are characteristic of their species, for the females remain silent until they are close enough to a male to "tch-tch"—advertise their presence. It is up to the female, however, to recognize the songs of their own males above the singing of other katydids, crickets, frogs, and all the other nocturnal vocalists—quite an accomplishment on a warm tropical night. The katydid in this illustration is an *Amblycorypha* from Washington, D.C.

47. This katydid, *Markia hystrix,* from Venezuela, is virtually invisible on a background of lichens. Experiments have shown that cryptically colored nocturnal insects almost always select a suitable background on which to rest during the day. It is tempting to imply that insects choose their background, but they do not have such human qualities. The probable explanation is that on unsuitable backgrounds they are continuously stimulated into activity and only when it is appropriate will their instinctive responses allow them to remain motionless.

48. Flatidae are, as their name implies, flat and disc-shaped and have the strange habit of resting on twigs and small branches in a long line, all facing the same way. In one African species there are two color forms, one green and the other red, and they are alleged to cluster on vertical branches with the red forms beneath the green ones so as to resemble a partially opened spike of flowers. Though common in the tropics, very little research has been carried out on them, and relatively few can be reliably named. This plant hopper was photographed in Manaus, Amazonas.

49. This pair of *Enallagma* demonstrate the curious and unique mating habits of damselflies and dragonflies. Prior to courtship, the male deposits sperm in a pouch at the base of his abdomen, just behind his hind legs. Then a successfully courted female is gripped behind the head by the claspers at the tip of the male's abdomen and held tightly while she brings up the tip of her abdomen to draw off the sperm from the male's pouch. In this picture the female damselfly is on the left. Sometimes the pair may take flight still coupled and fly in tandem for several minutes.

50. The various species of dragonfly employ different methods to lay their eggs. Some climb down the stems of water plants and, encased in a bubble of air, attach their eggs to the submerged vegetation. One might think that this hovering libellulid was embedding its eggs in a lily leaf, but it is only washing them off its abdomen, a very common dragonfly practice, though one which is usually performed while the dragonfly is skimming over the surface. Immature dragonflies and damselflies are carnivorous, like their adults, but hunt their prey underwater, some of the larger species even catching tadpoles and small fish.

51 & 52. One of the confusing aspects of collecting dragonflies is that males and females of some species look very different. Plate 51 is the female and plate 52 the male of *Palpopleura lucia* (Libellulidae) from Ghana. Moreover, unless great care is taken while they dry, the brilliant and characteristic colors soon fade. Dragonflies used to be pinned and spread like butterflies and moths, but with some wingspans of more than seven inches they needed an excessive quantity of cabinet space. The popular technique now is to fold the wings above the body and place each dried specimen in a

3½-inch-by-5-inch transparent envelope, together with a card bearing the date of capture, locality, and other data and then file it in a card-index cabinet. This system works well for us now, but what would we do if the two-hundred-and-fifty-million-year-old specimens with a twenty-eight-inch wingspan were not extinct?

53. Dragonflies are distinguishable from damselflies by their heavier bodies, more rounded heads, and relatively larger eyes, each of which may contain as many as 30,000 individual lenses. They depend upon their excellent vision to catch prey in flight, which they do by drawing their three pairs of legs under their face to form a catching basket. They regularly establish territories, where the insect can be seen repeatedly patrolling the same pathways for many days. Though in their immature stages they are fresh-water inhabitants, adult dragonflies extend their search for food far beyond their breeding grounds. Between their aerial excursions after insects, and when the wind is blowing hard, dragonflies sun themselves in the fashion of this libellulid from Wau, New Guinea.

54. If only the color of this skimmer *(Sympetrum)* from Washington, D.C., were permanent, what beautiful jewelry it would make! Perhaps it is just as well that the colors are ephemeral; otherwise, we might be adding still more species to our endangered list.

55. Most, but not all, damselflies hold their wings above their bodies while resting, as demonstrated by this calopterygid (possibly *Umma*) from Ghana. The bodies of damselflies are so slender and fragile that museum specimens usually have a bristle inserted along their length to give support. In some cases the differences in color between the sexes is so dramatic that unless a pair is taken while mating, one would not believe that they were both members of the same species.

56. This leaf-footed bug, *Anisoscelis foliaceae*, from Peru has been known to scientists for nearly two hundred years. It feeds while fully exposed on shoots and fruit, and presumably gains its protection from the vile-smelling fluid it emits when handled. It has been suggested that the legs are expanded to provide an expendable target for a would-be predator, but one sees very few leaf-footed bugs with missing legs. Maybe the bright colors and distinctive legs enable experienced birds to recognize them easily and remember the nasty flavor of the one they once tried to eat.

57. Cotton stainers, which are various species of *Dysdercus*, are well named, as they are known throughout the tropics as potential cotton pests. The damage they cause is due partly to the bacterial rots which enter through their feeding punctures. Economically more serious is the destruction that occurs when the fungus *Nematospora* infects the developing bolls: the cotton lint is stained an unbleachable yellow. Here we see a mating pair of *Dysdercus ruficeps* from Peru.

58. This beautiful beetle, *Sternotomis regalis*, from Ghana, has probably spent a year or more as an ugly grub tunneling in tree trunks. What beautiful pins these insects would make if only they were not so brittle!

59. As one of Nature's works of art, the diamond beetle *(Entimus imperialis)* is legendary. Peter Parley writes of it as follows in *Tales of Animals*: "This beetle belongs to the weevil tribe, and its scientific denomination is the Imperial Weevil. It inhabits South America, chiefly Brazil, and is the most resplendently colored of all the insect class. The ground color of the wings is a coal black, with numerous parallel lines of sparkling [round identations], which are of a green gold color, highly brilliant, from minute reflecting scales, like the scales of a butterfly. . . . There is another rich and elegant species of this insect in India; where, however, it is so scarce, that the wing cases (and sometimes the whole insect), are set like a gem on rings, and worn by the great." Could he have been referring to the weevil illustrated in plate 38?

60. The beetle family Cerambycidae is remarkable not only for the jazzy splendor of some of its members, such as this *Cyriocrates zonator* from Borneo, but for the surprising fact that their grubs are the only insects that have evolved a digestive juice that can break down wood. All other wood-eating insects depend upon bacteria or other micro-organisms to do the job for them. Some insects actually eat the wood only to keep their intestinal fauna flourishing, for a portion of the micro-organisms themselves are digested by the insect and used as its principal food supply.

61. Stink bugs are well named, for only if insects are well protected can they afford to be as conspicuously colored as this *Pachycoris torridus* from San Salvador. This is one of the several Pentatomidae known to indulge in parental care. The female stands guard over her eggs and permits the hatchlings to hide under the edges of her body when danger threatens. Some pentatomids have been seen trying to fend off parasitic wasps approaching the egg cluster, and it seems that they are reasonably successful, for the incidence of parasitism is high only around the edge of the cluster, where they are least able to be protected.

62. The forester's attitude toward Cerambycidae is very different from that of the insect collector, for large wood-boring beetles like this *Machrochenus tigrinus* from Sikkim can turn a tree trunk into a sponge in a matter of months. However, some experts believe that these beetles attack only unhealthy trees which are destined to die soon.

63. *Graphocephala coccinea* is a North American insect which has spread to Europe as a pest of rhododendrons, azaleas, and forsythia. It is perhaps one of the most spectacular leafhoppers (Cicadellidae), proving, once again, that an insect does not have to be rare in order to be beautiful.

64. Though this *Dysdercus andreae* was photographed in Florida, it is a migrant from the Greater and Lesser Antilles, where it feeds on cotton and a wide range of wild plants in the same family (Malvaceae). The over-all red color and black cross (hence its common name, St. Andrew's cotton stainer) is much more typical for *Dysdercus* than the three-tone species illustrated in plate 57.

65. Though most stink bugs feed on plants, there are a few which are predatory, like *Perillus bioculatus* from the United States and Canada, which feed on insect eggs, caterpillars, and beetle grubs exposed on herbs and shrubs. The temperature at which the immature bugs develop seems to influence the color and extent of the black areas of the adults, an inconsistency that has generated much interest among amateur collectors.

This species was deliberately introduced into France in the 1930s to control the Colorado potato beetle which had arrived by accident about 1870 and had become a serious pest. When the bug encounters a caterpillar or grub, it spears the victim with its beak and, holding it aloft, sucks out the body fluids, then casts away the empty skin. During the course of its development a single bug has been known to eat 450 eggs and 150 larvae.

66. This specimen of *Lophacris gloriosa* from Panama has been dried and posed in a position favored by museum workers. All the body components are spread apart, so that each aspect of the insect can be examined with ease. Grasshoppers are particularly prone to fade. This specimen was much greener in color during its lifetime. The red and purple hind wings probably enable the sexes to recognize each other visually from a distance, for they are weak singers. However, in some species, vivid areas such as these are flashed at overinquisitive birds with a quick separation of the folded wings. Once startled, the insect flies to a new retreat where it may find safety.

67. This Arizonian treehopper, *Platycotis tuberculata*, is a typical member of the temperate Membracidae, in that it does not show the extraordinary developments of the pronotum that are characteristic of the South and Central American species. Why the tropics should favor these unusual shapes is not known (see plates 15, 16, 33, and 35).

68. Moths in the family Ctenuchidae, of which *Histiaea meldolae* from Venezuela is an example, are brilliantly colored day fliers. Some species mimic wasps and other stinging insects. It seems more than likely that these species are distasteful to birds and other predators—otherwise they might not take such risks. In our studies with distasteful butterflies we have found that the noxious properties were originally acquired by the caterpillars while feeding on plants with poisons in their leaves. These toxins are stored during the pupal resting stage and later used by the adult to deter enemy attacks.

69. The arctiid moth *Melese*, from Venezuela, has been used experimentally to discover the function of the sound-producing organs on the sides of the thorax (see note to plate 18). In one experiment beetle grubs were projected into the air in a darkened room in which a wild-caught bat was free to fly. It was found that if a recording of the sounds produced by *Melese* was broadcast as the grub was flipped up, the bat would miss it, whereas without the recording, the grub would almost always be scooped up by the mouth or feet before it fell back to the ground. At first it was thought that the sounds jammed the bat's echo-locating device so it was unable to catch it, but the alternative view, now accepted, is that the message transmitted by the moth was, "I am distasteful, don't eat me."

The real moth is distasteful, and so, in the wild, bats soon learn to associate the sound with the bad taste and consequently avoid it. During the experiments the experienced bats heard the sounds and avoided the meal purposefully. Infrared film has been used to document and support the argument that the bats are deliberate in their avoidance, and that it is not just an error in aerobatics.

70. Nearly all the Cetoniinae beetles have a porcelaneous appearance, but few are so ceramic as this *Euploecila australiae* from Australia. These beetles are conspicuous during the day, as they sit on flower heads while chewing the nectaries. Their inelegant larvae are found in rotting vegetation.

71. In the summer of 1976 England seemed to be infested with ladybugs, or ladybirds, as they are known there. They swarmed everywhere; in some places it was almost impossible to walk across a grassy field without crushing thousands underfoot. On the beaches people even complained of being bitten by them—an unprecedented event. Nearly all ladybugs, including this *Neda norrisii* from Colombia, feed on aphids, both as larvae and as adults, so they are among the gardener's greatest allies. Presumably, an abundance of aphids in Britain had allowed the number of ladybugs to grow to astronomic size. But the pendulum will swing. Throughout 1977 a shortage of food for ladybugs has caused their population to fall far below normal levels, and 1978 should be a great year for the aphid.

72. Flies of the family Syrphidae, such as this *Toxomerus geminatus* from Washington, D.C., are superb wasp mimics, though they are without a sting and are completely harmless. They hover over flowers while they sip the nectar and are among the most expert aerial gymnasts. However, a close inspection reveals that they have only a single pair of wings—a sure test for a fly. Some immature syrphids are truly the gardener's friend, for as legless maggots they feed voraciously on aphids and other small insects.

73. Just as we know very little about the function of the huge jaws of the male Lucanidae (plates 22 and 23), so are we also ignorant of the function of the developments of the head in cetoniids like *Eudicella gralli* from South Africa. It has been suggested that these forked pikes are used in battling competing males, for it is true that these projections are well developed only in males. But, as is usually the case, we are long on speculation and short on observation.

74. It seems unbelievable that a larva resembling an anemic earthworm, that makes its living tunneling in decaying wood, could turn into a beetle as beautiful as this *Chrysochroa lepida* from Kilimanjaro in East Africa. Magnificent as the Buprestidae adults are, their larvae look like mere fish bait, though they are probably used as human food in some part of the world, for the larger species may reach a length of several inches. The larval diet is so poor that it may take them several years to attain adulthood. Some species, however, are serious pests of orchard trees and decorative timber trees such as the birch. They are probably the most spectacular family of beetles, and the wing covers are used by several primitive tribes as a source of material for making necklaces and other jewelry.

75. Little is known of the biology of *Comperocoris roehneri*, a predatory shield bug from Chile. Shield bugs are so called because of the shield-shaped triangular plate that covers the middle of their back. The family name Pentatomidae refers to the five joints that make up the antennae.

76. Almost fifty per cent of the native species of Buprestidae found in Australia belongs to the genus *Stigmodera*, which are not to be found anywhere else in the world. This species, *S. marginicolis*, with its striking pattern of red and black, is typical of the group. Buprestidae larvae tunnel in wood and, in this capacity, have traveled to many parts of the globe. One instance is known in Europe, where a North American buprestid emerged from a piece of wood that had been imported some twenty-five years earlier. Adults found in houses may have been attracted by the lights at night (which is the reason that they are sometimes known as fire beetles), but, of course, it is more than possible that they may have just crawled out of the woodwork.

77. A native North American bee—not the better-known honeybee— pollinating a *Zinnia multiflora* in Washington, D.C. When Columbus landed in 1492 there were no honeybees in the Americas, though by 1532 they had been introduced to Brazil and, not long afterward, were being cultivated in North America. The European honeybee is blind to the color red, so many of the native European flowers which depend upon the honeybee for pollination are blue or yellow. Little research has been carried out on North American bees, but the significance of the colors of native North American flowers is complicated by the presence of hummingbirds, which can clearly distinguish red from other colors.

78, 79, & 80. The wings of a fly *(Condostylus)*, a golden-eyed lacewing *(Chrysopa oculata)*, and a midge *(Chironomus plumosus)*, all taken in Washington, D.C., show the splendid effects that transparent wings can produce under suitable lighting. The wings themselves are virtually without pigment, though the lacewing appears very pale green in daylight.

81. Despite the brilliant colors, this is the wing of a moth, not a butterfly. However, as the bright colors suggest, *Chrysiridia madagascariensis* (alternatively known as *Urania ripheus*) flies by day, not by night. Moths of this family, Uraniidae, also occur in Central and South America and at certain times of the year undergo dispersal flights in great numbers. The caterpillars of the species illustrated feed on plants in the cactus-like family Euphorbiaceae, which is notorious for its poisonous sap. It is likely that the adults, too, are toxic if eaten.

The wings of this moth are still used extensively in the jewelry trade but were more important during the Victorian era. Allegedly, only insectary-reared specimens are currently used. This claim may be true, for keeping the pupae at different temperatures between 34 degrees centigrade and 41 degrees centigrade produces drastic differences in the wing colors of the adults. The resulting new colors could be useful to the trade.

82. *Campylotes desgodinsi* is a spectacular day-flying moth of the family Zygaenidae. Some other species in the genus have a wingspan of more than five inches.

83. The color patterns of Zygaenidae have intrigued lepidopterists for years. We now know that the extremely wide range of natural variation is due to both genetic factors (which determine whether the spots are red or yellow) and environmental influences, such as temperature and humidity (which affect the size and number of the spots). It is surely no coincidence that the burnets, as these zygaenids are called, show a relatively high rate of error in choosing their mates. Many instances are known where pairs of different species have been taken in copulation, an occurrence that is normally very rare outside the laboratory.

The fact that adult moths, such as this six-spot burnet *Zygaena filipendulae* from Europe, are poisonous protects them against bird attacks. Some species contain cyanide. However, it is not clear from where these poisons are obtained, for the caterpillars feed on leguminous weeds which are not known to contain any poisonous material. Presumably, the adult moths manufacture the toxins themselves.

On Photographing Insects

The main prerequisite for photographing insects is *patience*. I spend hours lying on my stomach or sitting in ponds observing their behavior so when the time comes to photograph I am able to anticipate their actions. I also choose early dawn and late afternoon as times to photograph because the cooler air during these periods slows down insect behavior. In addition, one is more likely to observe behavior patterns, such as feeding, cleaning, and mating, during the early morning.

As for equipment, I prefer to use a single-lens reflex camera because it allows me to look through the picture-taking lens and see the subject as it will appear on film. The lens I employ for about eighty per cent of my photographs is the 55mm Micro-Nikkor with which I can record a life-size image of the subject. Similar results could be achieved by using a close-up lens (a filter-type attachment which screws onto the front of the lens). This is the most economical approach, but the drawback is that the results will not be as sharp as those obtained by using a lens specially designed for close-up work. The other lens that I sometimes use is the Macro-focusing Vivitar zoom. While it does not give me as great a magnification of the subject as the 55mm Micro-Nikkor, it allows me to shoot from a little farther away from the subject.

When I am photographing extremely small insects, or details such as the scale formation of butterfly wings, I use Zeiss Luminar lenses, which come in focal lengths of 16mm, 25mm, 40mm, and 63mm. These lenses are made for especially large magnification. For over-all shots of insect habitats, I use 24mm and 28mm lenses. The 24mm lens can also be mounted backward on the camera to give a four-times magnification of the subject on film.

Normally I prefer Kodachrome 64 film, but when I'm shooting extreme close-ups I use Ektachrome, which gives greater color saturation.

I have taken many pleasing pictures by using natural light. In the field I often have an assistant hold a mirror to reflect sunlight onto the subject. The additional light brightens the shadow areas and gives better over-all color saturation. This procedure works well as long as the magnification of the subject is not too great. As the magnification increases, the depth of field (area of focus) becomes narrower. In order to get both the subject and the background in focus, the use of an electronic flash becomes essential. Using a flash unit also frees me from worrying about any blurring that may be caused by the movement of either the camera or the subject, because the short duration of the flash has the added benefit of freezing any motion.

K.B.S.

Suggested Reading

Borror, D. J., and R. E. White. *A Field Guide to the Insects of America North of Mexico*. Boston: Houghton Mifflin, 1970. One of the Petersen series of excellent field guides, containing general information on collecting techniques. (404 pp., 1300 illustrations, of which 142 are in color.)

Dethier, V. G. *To Know a Fly*. San Francisco: Holden-Day, 1962. A delightfully humorous account of the use of the fly in scientific investigation. (119 pp., illustrated with cartoons.)

Hutchins, R. E. *Insects*. London, Sydney, Toronto, New Delhi, Tokyo, and Englewood Cliffs: Prentice-Hall, 1966. A small but interesting account of some of the remarkable achievements of insects. (324 pp., 135 black-and-white photographs.)

Klots, A. B., and E. B. Klots. *Living Insects of the World*. London: Hamish Hamilton, 1975. A lavishly illustrated systematic account of the insects. (304 pp., 152 color and 139 black-and-white illustrations.)

_____. *1001 Questions Answered about Insects*. New York: Dover Publications, 1961. (260 pp., 31 photographs.)

Linsenmaier, W. *Insects of the World*. New York: McGraw-Hill, 1972. One of the most comprehensive books on insect natural history, replete with illustrations and information. (392 pp., 160 full-page color plates, over 200 line figures.)

Stanek, V. J. *The Pictorial Encyclopedia of Insects*. London, New York, Sydney, Toronto: Hamlyn, 1969. A 544-page book of more than 1000 black-and-white photographs, 48 full-page color plates, with extended and informative captions.

Zim, H. S., and C. Cottam. *A Guide to Familiar American Insects*. New York: Golden Press, 1951. An inexpensive but very useful field guide. (160 pp., 225 species illustrated in color.)

Index

(Numbers in italics are plate numbers.)